FLIGHT and MOTION

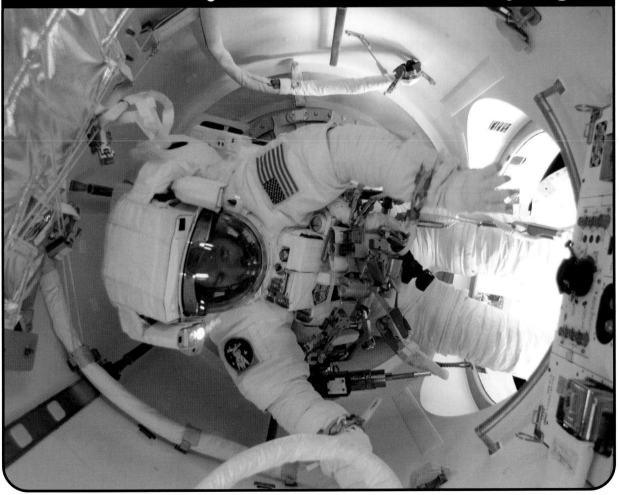

FLIGHT and MOTION

The History and Science of Flying

4

Mitchell – Space Probe

SHARPE REFERENCE
an imprint of M.E. Sharpe, Inc.

Sharpe Reference
Sharpe Reference is an imprint of M.E. Sharpe, Inc.

M.E. Sharpe, Inc.
80 Business Park Drive
Armonk, NY 10504

©2009 by M.E. Sharpe, Inc.

Library of Congress Cataloging-in-Publication Data

Flight and motion: the history and science of flying.
    v. cm.
  Includes bibliographical references and indexes.
   Contents: v. 1. Aerobatics–balloon – v. 2. Barnstorming–fuel – v. 3. Future of aviation–missile – v. 4. Mitchell–space probe – v. 5. Space race–Wright brothers.
   ISBN 978-0-7656-8100-3 (hardcover: alk. paper)
  1. Aeronautics–Encyclopedias. 2. Aeronautics–History–Encyclopedias. 3. Flight–Encyclopedias.

TL9.F62 2008
629.13–dc22

2007030815

CONTENTS

VOLUME 4

Mitchell, Billy

Date of birth: December 29, 1879.
Place of birth: Nice, France.
Died: February 19, 1936.
Major contributions: Led U.S. air forces in World War I; promoted the use of military aircraft.
Awards: Distinguished Service Cross; Distinguished Service Medal; Congressional Medal of Honor.

The son of a U.S. senator from Wisconsin, Billy Mitchell was 18 years old and still at college when the Spanish-American war began in 1898. He immediately volunteered for the army, entering as a private. His father used his influence to gain Mitchell a commission as an officer. Mitchell was assigned to the Signal Corps, the group that sent messages from one military unit to another. In combat, the young officer showed bravery and quick thinking.

World War I

Mitchell remained in the army after the Spanish-American war ended. As early as 1906—just three years after the Wright brothers took the first airplane into the sky—Mitchell predicted that future wars would be fought in the air. In 1912, by then a captain, Mitchell joined the Army General Staff as the youngest officer in that prestigious unit. While in Washington, D.C., Mitchell began his lifelong mission of urging the military to develop air power.

This photograph of Billy Mitchell with his U.S. Army plane was taken in 1920.

In his spare time, Mitchell learned to fly and gained his pilot's license. In 1915, he was assigned to the arm of the Signal Corps that was charged with developing a small air force. When the United States entered World War I in 1917, Mitchell was sent to France. He began talking to leading military figures from other nations allied with the United States who were interested in military aircraft. One of them was British general Hugh "Boom" Trenchard. The general argued strongly that air power should play an important role in allied operations. He is credited with advancing the

THE U.S. AIR SERVICE IN WORLD WAR I

When the United States entered World War I, its military air service was very small. The group numbered only 131 officers and about 1,000 enlisted men. It had fewer than 250 aircraft. The only manufacturing company in the country that could produce large quantities of planes belonged to Glenn Curtiss. He produced many of his famous "Jenny" training planes, and they helped the war effort. However, the United States did not produce a single combat airplane during the war.

role of military aircraft and in building Britain's Royal Air Force.

Mitchell agreed with Trenchard's ideas, and he went to work to create a U.S. air service. He began building airfields near places where American troops were stationed. Other officers often found Mitchell's personality to be brash and annoying, but he was determined to carry out his plan.

Mitchell was put in charge of all Allied aircraft during the Battle of St. Mihiel in September 1918. Mitchell commanded almost 1,500 planes—at the time, the largest air force ever assembled. In another battle later that fall, he sent massed forces of planes to carry out bombing missions.

Billy Mitchell's success in World War I led to his promotion to the rank of brigadier general. As second in command of the U.S. air service, he pushed for an independent air force.

Advocate of Air Power

Mitchell was promoted to the rank of brigadier general for his service in World War I. After the war, he returned to the United States as second in command of the air service. Mitchell urged research into better bombing sights, more powerful aircraft engines, and torpedoes that could be dropped by plane. He wanted to build planes that could carry troops and to form a separate air force with an independent command. He also managed to form an aerial force to fight forest fires.

Mitchell made sure that aviation stayed in the news and in the minds of Americans. He sent his pilots on speed and endurance flights to build publicity. In 1922, Lieutenant James Doolittle became the first person to fly across the United States in less than a day. The next year, Lieutenants John Macready and Oakley Kelly made headlines by flying across the country nonstop. In 1924, Mitchell sent eight airmen in four planes to fly around the world. Two of the planes crashed along the way, but two arrived back in Seattle, Washington State (their departure point), six months and 26,345 miles (42,389 kilometers) after taking off.

Opposition

Mitchell's work met resistance, however. Senior officers were not yet willing to accept the idea that air power would be important. They were outraged at Mitchell's charge that battleships had become outdated. At that time, U.S.

battleships were the largest and most powerful ships in any navy. Naval officers insisted that the defense of the United States depended on a fleet of these ships to block any invasion of the nation.

Mitchell countered that the ships could easily be destroyed by air. He campaigned in the press for the right to test his theories. He suggested a simulated attack on a German battleship seized at the end of World War I. In June and July 1921, Mitchell got his chance. In tests, as he had predicted, aircrews sank several ships, including four battleships. "No surface vessels can exist wherever air forces acting from land bases are able to attack them," Mitchell wrote.

Although proven correct, Mitchell remained unpopular in military circles. He continued to use the press to accuse senior military officers of ignoring air defenses. He toured U.S. naval bases in the Pacific Ocean and issued a stark warning: "If our warships [at Pearl Harbor, Hawaii] were to be found bottled up in a surprise attack from the air and our airplanes destroyed on the ground . . . it would break our backs. The same prediction applies to the Philippines."

Mitchell's words proved uncannily accurate years later, when the Japanese severely damaged U.S. ships and grounded airplanes with the 1941 attack on Pearl Harbor from the air.

↪ In one of Billy Mitchell's tests to prove the value of air power, an MB-2 aircraft successfully blew up an obsolete battleship in 1921.

In 1925, Billy Mitchell (standing) was court-martialed and found guilty of insubordination.

Court-Martial

In early 1925, Mitchell's appointment in the U.S. Air Service expired. Instead of renewing it, army commanders sent him to an isolated military base in Texas. Later that year, the navy suffered two air disasters when a seaplane broke down and a dirigible exploded. Mitchell immediately released a stinging attack on the heads of the navy and the army, accusing them of "almost treasonable negligence of our national defense."

His superiors had had enough, and they convened a court-martial. Mitchell was charged with insubordination (not obeying senior officers). After a seven-week trial he was found guilty. The verdict was suspension from duty for five years, but Mitchell decided to resign from the U.S. Army altogether.

Mitchell spent his remaining years writing and speaking to promote the ideas he had long advanced. He became ill in the mid-1930s and died at the age of fifty-six. During World War II, Mitchell's basic argument was proven true. Air power proved vital to Allied victory in both Europe and the Pacific.

In April 1942, a few months after the attack on Pearl Harbor, U.S. bombers attacked Japan using B-25s, nicknamed "Mitchells."

In 1946, ten years after his death, the U.S. Congress voted to award Mitchell a Congressional Medal of Honor, in tribute to foresight.

SEE ALSO:
- Aircraft, Military • Curtiss, Glenn
- World War I • World War II

Momentum

Momentum is a property of all moving objects, including both aircraft and spacecraft. An object's momentum is calculated by multiplying its mass by its velocity. Velocity is a vector—it has direction as well as size. Momentum, therefore, is also a vector.

A ball rolling down a hill gathers momentum as it goes faster and faster. A skydiver jumping out of a plane gathers momentum as gravity accelerates his or her rate of motion toward the ground.

Momentum depends on mass as well as speed, so a massive aircraft, such as a jumbo jet, has a lot more momentum than a smaller, lighter plane flying at the same speed. If an aircraft speeds up, its momentum increases. If it slows down, its momentum decreases. When it lands and comes to a stop, its momentum falls to zero.

The Law of Conservation of Momentum

When two or more objects exert forces on each other, their total momentum always stays the same. This is called the law of conservation of momentum, and it helps to explain why aircraft and rockets move.

A rocket engine sends out a high-speed jet of gas when it is fired. The rocket exerts a force on the gas and, according to Newton's third law of motion, the gas reacts by exerting an equal and opposite force on the rocket. The jet of gas has momentum in one direction. The only way that the total momentum of the rocket and gas can remain the same is if the rocket gains the same momentum in the opposite direction. So, the rocket moves. The same conservation law applies to aircraft. The momentum of the gas rushing out of an aircraft's jet engines is equal and opposite to the plane's momentum.

Angular Momentum

The momentum of an aircraft or spacecraft traveling in a straight line is called linear momentum. The momentum of something that spins is called angular momentum. An object's total angular momentum stays the same if no other

⟳ Marbles on a slope gather momentum as they roll. Objects with higher mass have more momentum.

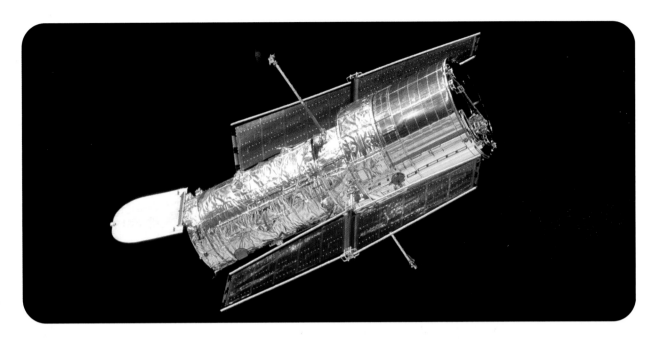

The Hubble Space Telescope is steered and steadied by momentum wheels.

forces act on it. This is also known as the law of conservation of angular momentum. It can be used to control the movement of a satellite in space.

The direction in which a satellite points is known as its attitude. Devices called momentum wheels, or reaction wheels, are often used to control a satellite's attitude. Inside the satellite, three wheels at right angles to each other are spun by motors. If a wheel is made to spin in one direction, then—to conserve angular momentum—the spacecraft must spin in the opposite direction.

The advantage of using momentum wheels to turn a satellite is that they use no fuel. If rocket thrusters or gas jets were used for attitude control, the satellite could only be controlled for as long as the fuel or gas lasted. After that time, the satellite would be out of control.

Momentum wheels should keep working as long as the satellite because they are powered by electricity generated from sunlight by solar panels.

Momentum wheels can turn a satellite to point at a precise part of the sky. The Hubble Space Telescope can take several hours to make an image of a very distant star or galaxy. It must point in exactly the same direction, without wavering, to capture a sharp image while it orbits Earth. The telescope uses momentum wheels to achieve this. Most Earth-orbiting satellites use momentum wheels to keep their antennae, solar panels, cameras, and other instruments pointing in the right direction.

> **SEE ALSO:**
> • Force • Gravity • Hubble Space Telescope • Laws of Motion • Rocket • Satellite •Speed

Montgolfier, Jacques-Étienne and Joseph-Michel

Dates of birth: Joseph-Michel: August 26, 1740; Jacques-Étienne: January 6, 1745.
Places of birth: Joseph-Michel: Blaruc-les-Bains, France; Jacques-Étienne: Annonay, France.
Died: Joseph-Michel: June 26, 1810; Jacques-Étienne: August 2, 1799.
Major contributions: Invented the first successful lighter-than-air balloon; carried out first flight of a manned balloon.
Awards: Order of Saint Michel.

The Montgolfier brothers were the sons of a prosperous paper manufacturer from southern France.

🎧 This double portrait of the Montgolfier brothers was based on a relief sculpture of their profiles made by Jean-Antoine Houdon.

Joseph-Michel was interested in science and had few business skills. Jacques-Étienne was trained to take over the family business.

During the early 1780s, Joseph-Michel noticed that a piece of paper rose in the hot air of a chimney. He started thinking about building a device that could use this effect to rise into the sky. Joseph-Michel built a small box of lightweight wood and covered it with fabric. He placed a wad of paper underneath the container and set the paper on fire. The container rose in the air until it hit the ceiling.

Joseph-Michel showed this discovery to his brother, and they began experimenting. In December 1782 the Montgolfiers built a larger container and tested it outside. The object rose nearly 70 feet (21 meters) on heated air and stayed aloft for a minute or so. The Montgolfier brothers went on to have more successful experiments. They built two larger globe-shaped containers held down by long ropes. The globes rose when heated, although they did not fly because of the ropes.

The Montgolfiers then prepared a large balloon to show the public their invention. They made a sphere out of tough sackcloth, covered it with four fabric panels held together by buttons, and surrounded the entire structure with netting to keep everything in place. On June 4, 1783, the Montgolfiers brought their invention to a square in their hometown of Annonay and lit a fire underneath the balloon. As a crowd

➲ On September 19, 1783, the Montgolfier balloon rose over the French royal palace of Versailles, watched by the king and queen and a large crowd of spectators.

watched in amazement, the balloon rose. The brothers estimated that it reached 6,000 feet (1.8 kilometers). Carried by the wind, it traveled about 1.2 miles (2 kilometers) before coming back to land.

The Montgolfiers took their discovery to Paris, where Jacques-Étienne demonstrated another balloon for members of the Academy of Sciences in mid-September 1783. King Louis XVI and Queen Marie Antoinette asked to see the invention for themselves. Accordingly, on September 19, 1783, Jacques-Étienne sent another balloon aloft. This time he attached a basket that carried a sheep, a rooster, and a duck. The balloon rose and floated on air for nearly 10 minutes, landing more than 1 mile (1.6 kilometers) away.

The first person to reach the balloon after it landed was a physician, Jean-François Pilâtre de Rozier. Excited, he volunteered to go aloft. During the next few weeks, Jacques-Étienne built a balloon that measured 75 feet (23 meters) high and 46 feet (14 meters) across. It had a basket with a container that could hold and sustain a fire to keep the balloon aloft. The brothers tested it several times with Jacques-Étienne or de Rozier. on board but with the balloon held down by ropes. Then, on November 1, 1783, de Rozier and a French army major entered the basket and lit the fire. Once the balloon was cut loose, it rose into the air. They cruised for 25 minutes, traveling about 5 miles (8 kilometers) over Paris.

The brothers continued experimenting. On January 19, 1784, Joseph-Michel went aloft along with de Rozier and five others. Thereafter, the Montgolfiers abandoned balloons and devoted themselves to other work. Joseph-Michel tinkered with some experiments not connected to flying. Jacques-Étienne returned to papermaking and invented a method for making vellum, a strong kind of paper.

SEE ALSO:
- Aerodynamics • Aeronautics
- Balloon

Myths and Legends

Humans have long been fascinated with flight, as ancient myths and legends reveal. Some stories were about flying gods, while others told of magical winged creatures. Some of the most interesting tales told of bold humans who attempted to fly.

Gods of the Air

It is not very surprising that ancient humans—bound to Earth—would believe their gods had the ability to fly. This ability underscored the difference between the human and the divine.

Some gods simply moved through the air in unexplained ways. Others are shown with wings. The ancient Egyptian god Horus had the body—or sometimes just the head—of a falcon. Garuda, a god of ancient India, had the body and arms of a human and the wings, head, and claws of an eagle. The Zoroastrian religion of ancient Persia showed its chief god, Ahura Mazda, with large wings—in some depictions, in fact, the god is just a circle flanked by huge wings. The Greek gods Eros and Nike were also shown with wings.

While many ancient gods could fly, this ability is generally not a central feature of myths about them. An exception is Helios, the Greek god of the Sun. He rode a chariot through the sky each day, carrying the Sun on its daily journey.

One interesting Greek myth tells of the dangers of misusing the power to fly, even for the gods. One day, Phaeton, the

The Egyptian god Horus was depicted as a falcon. People of ancient civilizations believed many of their gods had the ability to fly.

son of Helios, drove his father's golden chariot on his own. The young god did not have the strength or experience of his father, however, and the horses pulling the chariot went out of control. Zeus, king of the gods, knew that Earth would be scorched if this continued. He threw a thunderbolt to stop the runaway chariot. The horses returned to their normal course, saving humankind, but the young god was killed.

There are other ancient mythologies that include flying vehicles used by the gods. Baal, the god of the Canaanite people who lived in what is now Israel,

398

rode a chariot of clouds. Indian myths are full of stories about flying vehicles called vimanas, which the Hindu gods often used in battle.

Creatures with Wings

Along with gods and goddesses, mythologies are full of flying creatures, such as the winged horse Pegasus. Dragons are found in legends across the world, and many cultures depict them as having the power to fly.

The Oni of Japan were humanlike flying demons who used their sharp claws to take hold of the spirits of dying people who had led evil lives. Ireland had evil spirits that traveled on the west wind to grab the souls of the dying. Vampires also may fly, either on their own or by turning themselves into bats, depending on the version of the legend.

Some flying creatures were part human and part spirit. The Smaj of Serbia served as the protectors of the Serbian people and could spit fire on enemies from the air. The Kanae of Polynesia changed into flying fish, giving them the ability to travel through water or air.

The Jewish and Christian faiths also include creatures that can fly. The good ones, of course, are angels. The evil ones are demons, who carry out the work of Satan. Not all accounts of these creatures give them wings, but they were typically shown that way from the Middle Ages onward.

The mythology of Malaysia provides a different twist. It tells the tale of

PEGASUS AND BELLEROPHON

Bellerophon, a figure of Greek mythology, was a skilled horseman. During his travels, he was given the difficult task of fighting the Chimera, a monstrous beast that was part lion and part dragon. Acting on the suggestion that he use the winged horse Pegasus, Bellerophon placed a golden bridle—given to him by the goddess Athena—on the steed, thereby taming him. Mounted on Pegasus, Bellerophon killed the Chimera.

Bellerophon lived happily for many years until he decided to take another ride on Pegasus. Foolishly, he set his goal as Mount Olympus, home to the gods. Enraged by Bellerophon's boldness, Zeus sent a fly that bit Pegasus, causing the horse to buck. The sudden move threw Bellerophon from his mount, and he fell toward Earth. Athena prepared a soft landing for him on the ground, preventing his death. Bellerophon was left crippled, however, and Pegasus flew away.

🎧 This tomb in Turkey is the place where Bellerophon is supposedly buried.

Sheikh Ali, an evil ruler who controlled three armies of flying animals—horses, lions, and elephants.

Flying People

Stories about people who fly appear in cultures across the world. People of East Africa imagined Kibaga, a warrior hero, who soared over his enemies, dropping rocks on their heads. The Incas of ancient Peru told of Ayar Utso, who grew wings and flew up to the Sun. Koroglu, of Azerbaijan, mounted a horse that grew wings and carried him into the sky. Gatutkaca, from Java, had a magical jacket that allowed him to fly.

Perhaps the most famous myth about flying is the ancient Greek story of Daedalus and his son, Icarus. Daedalus had been brought by King Minos to the

🎧 The myth of Icarus is one of the most enduring stories of people's attempts to fly. This print shows Icarus falling from the sky after his wax wings melted in the heat of the Sun.

island of Crete to work for him. After Daedalus helped Minos's rival, Theseus, Daedalus knew that he and his son were in danger. For years, the two had secretly been making two pairs of wings out of eagle feathers attached to reeds by string and beeswax. Although the wings had not yet been tested, Daedalus and Icarus decided to use them to escape.

Before setting out, Daedalus warned his son not to fly too close to the sea, lest the spray of the waves wet the wings and cause him to fall. He also cautioned Icarus not to soar too high, which would allow the Sun's heat to melt the wax and destroy the wings. With those warnings, father and son put on the wings, climbed to a tower, and launched themselves.

The two soared over the sea, but Icarus was soon overcome with the thrill of flying. Forgetting his father's cautions, he began to climb higher and higher. Just as Daedalus had warned, the Sun's heat melted the wax holding his wings together. As the feathers fell off his arms, Icarus plunged into the sea and drowned.

The Greek hero Perseus fared somewhat better. His quest was to kill Medusa, a monstrous woman with

⟲ A colorful Indian carving shows Garuda, the flying carrier of the god Vishnu. In Hindu mythology, Garuda wears a crown and is huge enough to block out the Sun.

JATAYU AND SAMPAATI

Hindu mythology has a story similar to that of Daedalus and Icarus. The brothers Jatayu and Sampaati were the children of the flying god Garuda. Half-gods, they had the form of vultures. The two brothers competed to see who could fly the highest. One day Jatayu soared higher than his older brother. Sampaati feared that the Sun would scorch his brother's wings. He climbed above Jatayu to protect him. Jatayu was saved, but Sampaati's wings were scorched off his back, causing him to fall to Earth.

SANTA CLAUS

Based on the story of a Christian saint, the legend of Santa Claus draws on elements from Holland, Germany, and Russia. Although this legend is hundreds of years old, Santa Claus's ability to fly is rather recent. In 1823 an American named Clement Moore published a poem called "A Visit from Saint Nicholas." In the poem, Moore described how reindeer fly through the sky to take Santa from rooftop to rooftop, enabling him to reach every house. This airborne means of travel has become an important part of the Santa Claus myth.

By the mid-1800s, Santa Claus, depicted here on an 1868 box of candy, had become part of the American Christmas tradition.

gods, Perseus obtained a pair of winged sandals that he used to fly to Medusa's lair. He also borrowed the helmet of Hades, the god of the underworld, which made him invisible. With that ability, he could approach the monster unseen. As he arrived, Perseus looked at the monster's reflection in his shield, rather than directly at her face. Thus protected from her power, he was able to slay her.

Perseus did not suffer any evil consequences as a result of his flying. Unlike Bellerophon, he did not risk the gods' anger. Instead, he gave the winged sandals to the god Hermes.

The German legend of Wieland features an ironworker who fell in love with a beautiful swan-maiden. They married and lived happily for a time, but the swan-maiden yearned again for the freedom of flight and left him. Wieland was captured by an evil king who forced him to make weapons and other goods. Eventually, Wieland fashioned a suit with wings, which he used to escape from the evil king's dungeon. After killing the king with arrows and gaining his revenge, he fulfilled his other goal by joining his swan-wife in the sky.

Kings and Emperors

Kings and emperors often appear in legends about flying. The earliest known story of a flying person—about 4,500 years old—is the legend of King Etana of Sumer in Mesopotamia (modern-day Iraq). The king and his wife were not able to have a child, and he desperately wanted an heir. Following the instruc-

snakes for hair who was so ugly and evil that anyone who looked at her face turned to stone. With help from the

tions of the sun god, he freed a captured eagle. The bird carried King Etana to heaven, where he begged the goddess Ishtar for a child. She gave him a plant that both he and his wife ate, and the treatment worked.

Nearly as old is the Chinese legend of the emperor Shun. He used two oversized hats to fly. Once he employed this device to escape a burning tower. On another occasion, he used it to fly around his empire.

The Persians also told of a king who flew. The vehicle that King Kai Kawus used was of ingenious design. Workers attached long poles to the four corners of this throne. They tied meat to the top of each pole, and at the bottom of each pole they chained an eagle. When the eagles grew hungry, they beat their wings in an effort to reach the meat. That motion carried the throne aloft. This method worked, and the eagles carried the king into the sky. Unfortunately, they grew tired and stopped flapping their wings. When that happened, the throne tumbled to the ground.

A similar story involves the Greek conqueror Alexander the Great. He tied hungry griffins to poles attached to his throne. Griffins were half lion and half eagle. Alexander's story ends with a more direct moral than that of Kai Kawus, however. His vehicle stayed in the air for a week and brought him near the heavens. An angel then appeared and asked him why he wanted to see the

This stone carving of a griffin is on the fourth-century B.C.E. Temple of Apollo in Didyma, Greece.

heavens when he did not yet understand everything about life on Earth. Humbled, the conqueror returned to land.

Britain also has an ancient legend of a king who flew. King Bladud, who reigned in the ninth century B.C.E., had great intelligence and practiced magic. He fashioned a pair of feathered wings and launched himself into the air.

However, the king's flight ended in disaster. In some versions of the story, he plunged to his death. In others, he slammed into a wall. Either way, he lost his life and his kingdom, which was then inherited by his son—Lear. King Lear then became the subject of another legend, which was immortalized in a tragic play by William Shakespeare.

One Thousand and One Nights

A classic collection of stories from medieval times is *One Thousand and One Nights*. These tales from Southwest Asia relate the adventures of kings and councillors, fishermen and merchants, soldiers and slaves. In this world of magic and mystery, some stories involve that age-old dream of humans flying.

↩ By the 1800s and 1900s, science fiction had replaced ancient myths and legends about flight. Author Jules Verne described a journey to the Moon and back in *From the Earth to the Moon* (1865). The launch of Verne's fictional craft (illustrated here) took place in Florida, which later became the real launch site for the U.S. space program.

In the tale of an enchanted horse, a Hindu man crafted a wooden horse that could fly. A sultan's son used the horse to fly, whereupon he met a beautiful princess and fell in love. The princess ended up being the captive of another ruler, and the prince used the horse once again to save her.

In his second of seven voyages, the famous sailor Sinbad flew. He did so by tricking large birds to carry him out of difficult spots. In the course of this adventure, he also gained a fortune in diamonds.

A third tale from this legendary collection involves a magic carpet. Three princes all hoped to marry the same young woman. They agreed to a proposal that the one who brought back the most unusual gift would win her hand. One of them found a magic carpet in a market and bought it. The three princes used the carpet to fly to the woman's rescue, saving her life. In the end, however, the prince who had bought the carpet became a holy man instead of marrying the young woman.

SEE ALSO:
• Bird • Wing

MODERN MYTHS

Myths from many different cultures tell of gods who come down to Earth to meet with humans. Some people claim that these stories reveal visits from space travelers in ancient times and that some ancient drawings show gods in spaceships or wearing helmets. Scientists dismiss these claims, however. Today, stories about aliens from other planets focus on unidentified flying objects, or UFOs. Since UFO sightings in Washington and Idaho gained great media attention in 1947, sightings of UFOs have increased dramatically. By the 1950s, some people were beginning to connect UFOs with religious and supernatural beliefs. Claims of UFO sightings are most common in the United States. The kinds of UFOs people report most frequently are flying saucers or moving lights.

➲ A photograph from the files of the Central Intelligence Agency (CIA) shows what the photographer claimed was a UFO over New Jersey in 1952. Many UFO images appeared in the period, and there was much doubt about the authenticity of the images.

NASA

The National Aeronautics and Space Administration (NASA) is the U.S. national space agency. It was formed in 1958 for advanced aeronautics research and space exploration. NASA is a federally funded organization, employing thousands of engineers, scientists, and professionals in aerospace research. Its work includes developing new airplanes and spacecraft and testing new technologies.

NASA has been associated with many of the most dramatic and historic episodes in the history of spaceflight—it achieved worldwide recognition in the 1960s when it sent astronauts to the Moon. NASA's work continues into the twenty-first century, with manned spaceflights and with space probes that explore the Solar System. It is the world's leading space agency, ahead of the Russian federal space agency and the European space agency.

NASA scientists and engineers are also engaged in research projects concerning transportation and the environment. Images taken from NASA sources, such as space telescopes and probes, have excited the imaginations of people around the world. NASA's extensive educational and media programs provide information about space and space technology.

NACA

The predecessor to NASA was known as the National Advisory Committee for Aeronautics (NACA), which was founded in 1915. This body was responsible for important early research into airplane flight, using research airplanes and wind tunnels. By modern standards NACA was small—in 1938 it had a staff of just over 400 people.

After World War II (1939–1945), NACA expanded its activities into the realm of supersonic flight, working closely with the U.S. Air Force on the record-breaking X-1 airplane and other projects. In the late 1940s, the Department of Defense urged scientists to work with the military on missile experiments. At the same time, scientists were pressing for rockets to be sent into space for research. President Dwight D. Eisenhower approved a plan to launch a science satellite as part of the International Geophysical Year, scheduled for July 1957 to December 1958. The chosen rocket vehicle for the satellite launch was the Naval Research Laboratory's Vanguard rocket.

The Vanguard Project was underfunded and slow to get off the ground. The United States was shocked when, in October 1957, news broke that the Soviet Union had beaten America into space by launching the world's first artificial satellite, *Sputnik 1*. Many people in the United States became concerned that there was a widening gap between Soviet and U.S. space science. American scientists quickly responded to the challenge, launching the nation's first satellite, *Explorer 1*, in January 1958. Despite this achievement, however, there were calls for a new agency to drive forward the national effort in the "space race."

A New Agency

On October 1, 1958, Congress created a new organization "to provide for research into the problems of flight within and outside the Earth's atmosphere, and for other purposes." This new organization was the National Aeronautics and Space Administration, or NASA.

NASA had broad goals linked to the needs of national defense and the advancement of U.S. space science. It was hoped, through the direction of a single agency, that NASA would avoid the duplication of effort that had occurred through separate U.S. Air Force, Army, and Navy rocket programs.

When NASA came into being on October 1, 1958, it absorbed NACA's employees (there were by then 8,000 of them) and its three major research laboratories: Langley, Ames, and Lewis. NASA also acquired the facilities operated by the Jet Propulsion Laboratory (JPL). This lab, run by the California Institute of Technology for the U.S. Army and the U.S. Army Ballistic Missile Agency, was where rocket pioneer Wernher von Braun and other engineers were at work on long-range missiles.

⏺ The drafting room at the NACA Airplane Engine Research Laboratory in the early days was a long way from the high-tech NASA facilities of today. The laboratory has since become the Langley Research Center in Hampton, Virginia.

NASA'S LEADERS

The head of NASA in 2008 was administrator Michael Griffin. His distinguished predecessors included Robert Gilruth, head of NASA's Manned Spacecraft Center until 1973, who did much to ensure the success of the Apollo program. Another eminent name in NASA history is that of Chris Kraft, NASA's first flight director. He succeeded Gilruth as head of the Manned Spacecraft Center. NASA's expertise in unmanned, long-distance space exploration owes much to Bill Pickering. Originally from New Zealand, Pickering became a U.S. citizen in 1941 and worked on the first U.S. satellite at the Jet Propulsion Laboratory.

◗ During his forty-year career with NASA and its predecessor NACA, Robert Gilruth (1913–2000) led many U.S. spaceflight operations.

Projects Mercury and Gemini

NASA quickly captured the public imagination with Project Mercury. Amid a blaze of publicity, seven pilots (all men) were chosen to be America's first astronauts. Alan B. Shepard was the first American to fly into space on May 5, 1961, squeezed inside a cramped Mercury capsule launched by a Redstone rocket. On February 20, 1962, John H. Glenn, became the first U.S. astronaut to orbit Earth.

As the Mercury program continued, NASA scientists also were engaged in a range of unmanned space activities—sending probes to the Moon and to Mars, for example. Public attention, however, focused on the "space race" between the Soviet Union and the United States. The declared U.S. intention, as stated in May 1961 by President John F. Kennedy, was to land men on the Moon and bring them back safely. This was an immense challenge, and many people doubted NASA could achieve the president's goal.

After the completion of the Mercury program, NASA progressed to two-person flights in Earth's orbit, using the larger Gemini spacecraft. Gemini flights provided valuable experience in space piloting, rendezvous and docking, extra-vehicular activity (space walks), and reentry and splashdown techniques.

President John F. Kennedy (center, facing right and wearing sunglasses) toured NASA's facilities at Cape Canaveral in Florida in 1962. The Cape was the launch site of the Apollo missions. Today, NASA's facilities at the Cape and elsewhere have expanded greatly.

Apollo

The Moon landing program, using the three-person Apollo spacecraft, was pursued with enormous energy at a staggering cost of over $25 billion. Apollo has been compared, in terms of national effort, to digging the Panama Canal or making the first atomic bomb during World War II. In 1967, the Apollo program survived the tragic setback of a fire inside an Apollo capsule in which three astronauts died. The program triumphed in 1969 with the historic landing of two Apollo 11 astronauts, Neil Armstrong and Buzz Aldrin, on the Moon.

Everything NASA did was very public. During the Apollo 11 spaceflight and later Apollo Moon landings for example, people all over the world were enthralled by television coverage of the launches and splashdowns, directed from NASA nerve centers that included the Cape Canaveral launch site in Florida and the Mission Control Center in Houston, Texas. Television audiences were able to see control room staff at work and talking to the astronauts, and they could watch pictures beamed directly from the Moon. NASA space jargon used by the flight controllers, such as "T minus 30 and counting" have since passed into common usage.

Five more lunar landings followed that of Apollo 11. NASA managed to avert disaster when the Apollo 13 mission of April 1970 went seriously wrong. On this mission, the Moon landing had to be canceled after an oxygen tank exploded midway through the outward flight. The three astronauts flew around the Moon and, despite severe power problems, returned safely to Earth. Their safe return was a tribute to NASA's ability to adapt its technology to cope with the unexpected. In total, twelve astronauts walked on the Moon during the

🎧 NASA workers at Mission Control in Houston, Texas, celebrate the successful conclusion of the Apollo 11 mission in 1969.

six Apollo lunar landings, which marked the highpoint of NASA's success.

The ISS and Space Shuttle

After the Apollo mission, NASA experienced a falling off in public interest in space. The agency also was hampered by financial restraints, and it had to cut back on some programs. In 1975, NASA cooperated with the Soviet space agency to run the Apollo Soyuz Test Project, a joint flight by U.S. and Soviet astronauts. The project foreshadowed today's cooperation with Russia and other

nations on the International Space Station (ISS). NASA had originally planned to launch its own space station, as authorized by Congress in 1984. Eventually, with costs high and rising, it was decided that an international partnership was more appropriate. The ISS was the successful result of this cooperation, and it has been consistently crewed by a changing group of astronauts since 2000.

In 1981, NASA astronauts flew the first reusable Space Shuttle; in 1983, NASA astronaut Sally K. Ride became America's first woman in space when she flew on the STS-7 Space Shuttle mission. The Space Shuttle, used to supply the ISS and for satellite launches and other duties, has absorbed much of NASA funding and posed some major challenges since it first flew. It has proved a valuable spacecraft, however.

Two major Space Shuttle accidents caused some critics to question NASA's safety standards and operational systems. In 1986, *Challenger* exploded shortly after takeoff, killing all seven crew members. In 2003, *Columbia* broke up shortly before it was due to land. Again, all seven crew members died. Space Shuttles were grounded after each of these disasters, while NASA and its partners involved in the program investigated the causes. In both cases, a fault was identified and rectified by design changes. The three remaining Space Shuttles were back in operation by 2005.

⬤ The International Space Station, a cooperative project involving several nations, began with the launches of Russian module *Zarya* and U.S. module *Unity*. This photograph shows *Unity* in the foreground, being delivered by NASA's Space Shuttle *Endeavour* for rendezvous with *Zarya*.

Going Farther

NASA has achieved an impressive record of exploring the Solar System and beyond with unmanned probes, satellites, and space telescopes. In the 1970s, *Pioneer 10* and *11* flew past Jupiter and Saturn. They were followed by *Voyager 1* and *2*, record-breaking probes that made a tour of the outer planets before eventually leaving the solar system. In 1976, NASA landed two Viking spacecraft on the surface of Mars, and these landers sent the first pictures from the surface of the red planet back to Earth.

Not even NASA's highly trained and experienced engineers are infallible, however. Sometimes spacecraft disappear. In 1993, the *Mars Observer* spacecraft disappeared from tracking screens just three days before it was scheduled to go into orbit around Mars. A successor spacecraft, *Mars Global Surveyor*, made it into orbit safely in 1998.

NASA often has proved itself adaptable to challenges. After the Hubble Space Telescope was launched in 1990, scientists discovered that it had a faulty mirror. NASA designed a rescue package to deal with the unexpected problem in such a costly piece of space hardware. The agency sent Shuttle astronauts to correct the fault, which they did, and Hubble began to provide Earth-based astronomers with their clearest view yet of the heavens.

NASA Today

NASA today has ten major centers around the nation. The Kennedy Space Center at Cape Canaveral, Florida, is probably the best known. The others are:

☞ Personnel from NASA's Jet Propulsion Laboratory prepare *Mars Global Surveyor* for transfer to the launch pad. NASA's success in exploring the solar system has greatly increased human knowledge of space.

Ames Research Center, Dryden Flight Research Center, Glenn Research Center, Goddard Space Flight Center, the Jet Propulsion Laboratory, Johnson Space Center, Langley Research Center, Marshall Space Flight Center, and Stennis Space Center. All NASA activities rely on teamwork, not only among personnel at the various centers, but also between NASA and its partners in industry and the academic world.

Today, space is a business. The NASA launch services program based at the Kennedy Space Center offers commercial launch services from a number of launch sites. The sites include Cape Canaveral Air Force Station in Florida; Vandenberg Air Force Base in California; Wallops Island in Virginia; Kwajalain Atoll in the

Republic of the Marshall Islands; and Kodiak Island, Alaska. To provide a range of launch options, NASA buys expendable launch vehicle (ELV) services from commercial providers—for example, Atlas rockets are built by Lockheed Martin and Deltas are built by Boeing. NASA also works closely with international partners. The *Cassini* spacecraft, for example, was developed by the Jet Propulsion Laboratory in association with the Italian space agency. Launched in 1997, *Cassini* arrived at Saturn in 2004.

NASA also carries out research into supersonic flight within the atmosphere, following up on the pioneer work done by NACA. In the 1960s, the record-breaking X-15 rocket plane soared so high and so fast that it almost became a spacecraft. Its flights provided valuable data and pilot experience for the manned space program. NASA continues to research high-speed flight in the atmosphere. In 2004, the X-43A scramjet set a new world speed record for an aircraft with an air-breathing engine, flying at ten times the speed of sound.

NASA's long-term ambitions for the twenty-first century include sending astronauts back to the Moon and designing a mission to explore Mars. The program will involve construction

The X-43A scramjet is suspended in the air for controlled radio frequency testing. The aircraft, part of NASA's hypersonic flight program, set a new flightspeed record in 2004.

of a new generation of spacecraft, including the Orion manned spacecraft. In addition, NASA will continue its ambitious scientific program of exploring the universe.

SEE ALSO:
- Apollo Program • Astronaut
- Cape Canaveral • Kennedy Space Center • Satellite • Spaceflight
- Space Race

NEEMO

One of the more unusual facilities used by NASA is located 62 feet (19 meters) underwater. To train astronauts for NASA Extreme Environment Mission Operations (NEEMO), NASA sends them to Aquarius, off Key Largo, Florida. Aquarius is an underwater laboratory belonging to the National Oceanic and Atmospheric Administration (NOAA). Here, humans can experience life in an artificial habitat similar in many respects to being in space. NASA crews have stayed in Aquarius for between two and three weeks to train for missions to the Moon. They test techniques for communication, navigation, geological sample retrieval, construction, and using remote-controlled robots. Facing these challenges in Aquarius helps NASA's designers and engineers improve designs of habitats, robots, and spacesuits for future lunar projects.

Astronauts in training pose for a photograph inside and outside NOAA's laboratory.

Navigation

Navigation is the steering or directing of a course. Migrating birds, animals, and even insects seem able to navigate across the world with ease. People have developed ways of using nature, science, and technology to do the same thing—to figure out their position and find their way across land, sea, sky, and even in space.

Following Instinct and Landmarks

Monarch butterflies fly more than 1,500 miles (2,400 kilometers) on their annual migration across North America. One seabird, the Arctic tern, makes the longest migration journeys of any living creature. Every year, it flies up to 22,000 miles (35,400 kilometers) between the Arctic and Antarctic. Some animals are born with an instinct for migrating in a particular direction. Birds may navigate by recognizing familiar landmarks such as rivers and mountains. They also may use the position of the Sun and stars. Yet others seem to be able to sense the Earth's magnetism, as if they have a natural compass that directs them.

The first pilots relied on navigation methods similar to those used by birds. Planes flew low so that pilots could navigate visually by following landmarks such as roads, rivers, and railroads. For longer flights and for flights over oceans, a method called dead reckoning was used. A pilot used a map to figure out which direction to fly and then

Monarch butterflies fly more than 1,500 miles (2,400 kilometers) on their annual migration across North America.

measured the distance to the destination. Knowing how fast a plane flew, a pilot could figure out the journey time. If the plane was flown in the right direction (using a compass) at the correct average speed for the calculated length of time, it should arrive at its destination. In the real world however, an aircraft could be blown off course by wind, so pilots had to allow for this when plotting their course. Today, pilots of small aircraft still can navigate using dead reckoning and by looking out for landmarks.

THE STARDUST MYSTERY,

In 1947, an airliner called *Stardust* was flying from Buenos Aires, Argentina, across the Andes mountain range to Santiago, Chile. Just before it was due to land, it vanished. Searchers found nothing. In 2000, the wreckage was found, and an explanation to the old mystery was pieced together. Because of bad weather, the airliner had flown so high that it reached the high-speed air current of the jet stream and was flying against it. The crew's navigation calculations indicated that they had crossed the mountains, but the jet stream had slowed them down so much that they were still over the mountains. Thick clouds prevented them from seeing the ground. As they descended to land, the plane crashed into a mountainside and fell onto a glacier, a slow-moving river of ice. The wreck was soon covered with snow and then sank into the glacier. It took fifty-three years for the wreckage to travel downhill inside the glacier and appear at the bottom.

🎧 Early airplanes had open cockpits and flew relatively low to the ground. These factors allowed pilots to navigate by visual landmarks in the days before electronic navigation systems.

Celestial Navigation

Until electronic navigation systems were developed, pilots also navigated during long flights by using the positions of the Sun, Moon, and stars as explorers on land and sea had done for centuries. Navigation of this kind is called celestial navigation. Using a device called a sextant, the positions of the Sun, Moon, or certain known stars can be measured to pinpoint an aircraft's location. Each sighting enables the navigator of an aircraft to draw a line on a map. After several sightings, the point at which the lines cross show the airplane's position.

Celestial navigation involves making repeated sightings, doing many calculations, and plotting maps. For this reason, the crews of long-range airliners and bombers using this system included a specially trained navigator.

Electronic Systems

A variety of electronic navigation aids have since been developed to help pilots navigate more accurately. The most basic are beacons, or radio transmitters,

THE COMPASS

The magnetic compass was developed in about the twelfth century in both China and Europe. People noticed that a type of rock called lodestone, when placed on a piece of wood floating in water, caused the wood to turn. It always turned so that one end pointed north. When a magnetic needle is used instead, the needle turns in the same way to line up with the Earth's magnetic field. In a simple magnetic compass, the needle is on a dial marked with points of the compass—north, south, east, and west. Because the needle always turns to point north, a person holding a compass can figure out which way to head if, for instance, he or she wants to go west.

A normal magnetic compass does not work well in an aircraft. It swings wildly when the aircraft turns, and it is inaccurate near Earth's poles. A different type of compass, called a gyrocompass, does not use magnetism. Instead, it uses a spinning wheel called a gyroscope that keeps pointing in the same direction. A gyrocompass installed in an aircraft keeps pointing north, whatever way the plane moves or turns, so it can be used as a reliable compass.

The MH-53J Pave Low III heavy-lift helicopter is the largest and most powerful used by the U.S. Air Force. It also has very advanced navigational abilities, with an inertial navigation system, terrain-following radar, forward-looking infrared sensors, GPS capability, and a projected map display.

on the ground. Their positions are marked on navigation maps. A radio in an aircraft picks up radio signals from the nearest beacons and figures out their bearings (the direction of the beacons in relation to the plane). Knowing this helps a pilot to determine an airplane's position and to steer an accurate course. A variety of systems use radio beacons, including NDB (non-directional beacons), VOR (VHF omnidirectional radio range) and LORAN (long-range navigation). There also is a more accurate version of VOR called TACAN (tactical air navigation) for military aircraft.

When computers became small enough and reliable enough to be carried by aircraft, much more advanced navigation systems became possible. One of these is the inertial navigation system (INS) or inertial guidance system (IGS). The location of an aircraft is programmed into the system at the beginning of a flight. As the plane flies along, the system detects its movements by using devices called accelerometers. Knowing how much the aircraft has moved, and in which direction, enables its computer to determine the aircraft's position and to keep track of its progress.

An inertial navigation system can navigate a plane automatically. The system is programmed with the locations during a flight where the plane has to turn. These places are called waypoints. During the flight, the inertial navigation system controls the plane's autopilot and flies the plane along the planned route from one waypoint to the next.

A modern airliner is equipped with several navigation systems so that if one should fail, another can take over. If everything fails, an air traffic controller on the ground can guide the pilot or pilots of an aircraft over the radio.

Satellite Navigation

The most advanced navigation system uses the Global Positioning System (GPS), a network of navigation satellites orbiting Earth. The GPS system carried by an aircraft picks up radio signals from at least four satellites and uses them to calculate the aircraft's position, altitude, heading, and ground speed. Space-based navigation systems like this are beginning to replace radio navigation systems because they are more accurate. In addition, they do not rely on large numbers of beacons on the ground; they are not affected by bad weather; and aircraft are never out of range of the system's signals.

Navigating Spacecraft

All the planets in the solar system are spinning as well as moving around the Sun at very high speeds. Navigating a space probe from Earth to another planet could be compared to sitting on a spinning merry-go-round and trying to throw a ball at a spinning top

⤺ The GPS control room at Schriever Air Force Base in Colorado controls the satellites that provide navigational data to users around the world.

placed on a distant moving car. In spite of the challenges, however, space scientists have figured out how to send spacecraft where they want them to go. Most of the work needed to guide a space probe is done before the launch.

The movements of all the planets are known, and scientists can predict exactly where they will be at a given point in the future. The timing of a probe's launch, the speed it travels, and its direction are all chosen so that the probe is launched from Earth on the right flight path to reach a planet. The pull of the Sun's gravity and that of the planets has to be taken into account when calculating the probe's flight path. In fact, gravity is sometimes used to accelerate a probe or to change its direction without having to burn any fuel.

When the Space Shuttle goes to the International Space Station, its launch time is chosen to place it in orbit near the Space Station. The Space Shuttle only has to make small adjustments to its position to rendezvous with the Space Station.

Navigating the Apollo Missions

The only manned spacecraft that have navigated through space and into an orbit other than Earth's were those of the Apollo mission, when nine spacecraft went into orbit around the Moon. The Apollo program also landed twelve astronauts on the Moon between 1969 and 1972. On the way to the Moon, the spacecraft's inertial guidance system figured out its position by sensing changes in its speed and direction. The position was double-checked by taking sightings of stars using a sextant and telescope. Thirty-seven stars were used as guides to navigate the spacecraft. The sextant, telescope, inertial guidance system, and guidance computer provided Apollo's primary guidance, navigation, and control system (PGNCS). The astronauts called it "Pings."

Timing and precision are crucial to space navigation. Scientists successfully launched the probe *Deep Impact* to intercept the comet Tempel 1 in 2005. The probe released an impactor to create a crater, so releasing debris to gain information about the comet's interior.

 Goldstone Deep Space Communications Complex in California is one of three centers for the Deep Space Network. The network has provided navigation for many spacecraft.

The most precise Apollo navigation system was not in the spacecraft at all. It was on Earth. The huge radio dishes of NASA's Deep Space Network were trained on the spacecraft and relayed communications between Apollo missions and Earth. Tracking a spacecraft using these dishes showed exactly where it was. This information was sent up to the spacecraft and used to correct any errors in its own guidance system. If the spacecraft lost radio contact with Earth, its own guidance system would be correct and could guide it until contact with Earth was restored. Some losses of contact were expected. While the spacecraft was in orbit around the Moon, it lost touch with Earth every time it disappeared around the far side of the Moon. NASA still uses the dishes of the Deep Space Network in California, Spain, and Australia to communicate with space probes in all parts of the solar system.

AN EXPENSIVE ERROR

When navigation goes wrong, the results can be catastrophic. In 1998 NASA sent a probe named the *Mars Climate Orbiter* to Mars. It was supposed to orbit Mars, but instead it entered the planet's atmosphere and burned up. An inquiry found that some of the navigation data was calculated using U.S. standard units (feet/pounds/seconds), and this became mixed up with data calculated in metric units (meters/kilograms/seconds). The navigational error resulted in the loss of the multimillion-dollar spacecraft.

SEE ALSO:
- Bird • Communication • Global Positioning System • Radar
- Space Probe

Newton, Isaac

Date of birth: December 25, 1642.
Place of birth: Woolsthorpe, England.
Died: March 20, 1727.
Major contribution: First formulated the idea of gravity; expressed three key laws of motion; invented calculus, a branch of mathematics.
Awards: Named a Fellow and later President of the Royal Society; Order of Knighthood.

Isaac Newton had a difficult childhood. His father died a few months before he was born. Newton was a sickly child who was not expected to live. When Newton was two, his mother remarried, and his stepfather sent him away to live with other relatives. Only when the stepfather died was Newton reunited with his mother.

Newton began studying at Trinity College, Cambridge, England, in 1661. He dedicated himself to new, emerging ideas in science and mathematics. Much of his study of the new science was done on his own and appeared only in his private notebooks. In 1665 Newton graduated from Trinity College and over the next two years, he reached conclusions that formed the basis of his later work.

In 1667, Newton returned to Cambridge as a professor. Using a pair of prisms, he broke white light into its different colors—those we see in a rainbow. Through this work, Newton proved that light is not a simple structure but a complex one. He also began to develop the

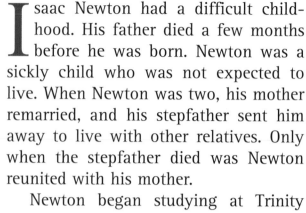
Isaac Newton is one of the most important scientific figures in the history of the world. His view of physics prevailed until they were modified by Albert Einstein's ideas in the early 1900s.

ideas that would become calculus, an advanced form of mathematics that makes it possible to describe the movements of objects.

Newton began to think about the orbit of objects around planets. He wrote an essay called *On Motion* in 1684 and expanded his ideas three years later in his masterwork *Philosophiae Naturalis Principia Mathematica*. With this work Newton established his three laws of motion and his theory of universal gravitation. Newton's ideas gained wide acceptance fairly quickly, and his theories are the basis of much of modern physics.

In 1696 Newton was named Warden of the Royal Mint, which oversees the making of coins. He kept his professorship at Cambridge for a few more years, but most of his later life was spent in London. In 1704, Newton finally published a work about sight, light, and color entitled *Opticks*.

Although he did little new work in later years, Newton remained an important figure in the sciences. He became President of the Royal Society in 1703. In that role he helped sponsor the work of younger scientists who quickly rose to dominate the scientific community in England, further helping to spread Newton's ideas. Two years later, Newton became the first scientist ever to be knighted by a British monarch.

Newton's later life was marred by a major controversy. He had devised the basics of calculus back in the 1660s but had never published a detailed

NEWTON'S CANNON

Hundreds of years before it actually happened, Newton understood that a human-made object could orbit Earth, just as the Moon did. Imagine, he said, a mountain so tall that it extended above Earth's atmosphere. Now imagine placing a cannon atop the mountain's peak and firing a cannonball on a horizontal path. A powerful enough cannon could fire the ball so fast that it would never fall. At the same time, gravity would bend its path toward Earth. Never falling and never escaping Earth's gravity, Newton theorized, the cannonball would orbit the planet indefinitely.

description. In 1684, the German mathematician Gottfried Wilhelm Leibniz published his work on calculus and gained renown as the inventor of a new system. Newton began a running battle with Leibniz that was carried out in print. Both men tarnished their names with their bitter attacks on one another. Newton died at the age of eighty-four.

SEE ALSO:
- Gravity • Laws of Motion
- Momentum • Satellite

Night Witches

Night Witches were a group of women combat pilots in the Soviet Union who fought in World War II (1939–1945). They flew at night, and they were so deadly that the Germans gave them the nickname of "Night Witches."

Women in Combat

The Soviet Union was the only Allied nation in World War II to use women pilots in combat. Women flew military aircraft in the United States, Canada, and the United Kingdom, but only on transport or delivery flights. Such missions could be difficult and often hazardous, but there was a low risk of encountering enemy aircraft.

During World War II, Soviet armed forces fought a desperate battle against large German armies following the German invasion of Russia in the summer of 1941. The Soviets were desperately short of planes and pilots. In the Communist Soviet Union, women drove farm tractors and trucks alongside men, so—Soviet generals reasoned—why not train them to fly combat planes?

Three regiments of female combat pilots were formed in the Soviet Union in 1942. The Night Witches belonged to the 588th Night Bomber Regiment. The pilots, mechanics, and weapons personnel in this unit were all women.

The Night Witches flew old-fashioned Polikarpov Po-2 biplanes, really only suitable as training airplanes. These wooden and fabric planes had a top speed of only 94 miles per hour (150 kilometers per hour). The Po-2 biplanes had such weak engines that they could carry only two bombs. Possibly female pilots were given such old planes because the military did not have great concerns for their safety, but the Night Witches developed some tricks that surprised everyone.

Night Witch Tactics

Mostly, the Night Witches made surprise raids on German supply depots and army camps. They flew by night, and the pilots often shut off their engines when approaching the enemy position so that the planes glided in silently. They restarted the engines just before the attack, dropped their bombs, and disappeared into the darkness before German troops could open fire.

◖ A group of female Soviet combat pilots make their flight plans somewhere in the Soviet Union in 1942.

🎧 Stalingrad in Russia was destroyed as German and Soviet forces fought in the skies and on the ground during the Battle of Stalingrad.

The slow Po-2 biplane appeared to be an easy target for a fast German fighter, such as an Me-109. The top speed of the Soviet biplane, however, was actually less than the minimum (stall) speed of the German fighter. This difference in speed meant that a German fighter chasing a Po-2 often sped past its slow-moving target and had to fly around in a circle for another attack. The skillful Night Witch pilots, meanwhile, flew close to the ground, twisting and turning and even disappearing behind trees. The German fighter pilots did not find it easy to shoot down a Night Witch.

The Russian planes were so small and flew so low that they barely showed up on German radar. In key battles, such as the Battle of Stalingrad in 1942–1943, the Germans mustered searchlights and used heavy antiaircraft guns to blast the Night Witches out of the sky. The pilots of the 588th Regiment flew in threes to outwit this tactic. Two planes flew straight and level, to attract the search-lights, and then they began a series of aerobatic moves. While the German searchlight crews struggled to hold their light beams on the gyrating biplanes, the third Night Witch moved in to attack. The pilots then regrouped and repeated the attack until all three aircraft had dropped their bombs.

Twenty-three Night Witch pilots received their country's highest honor in the form of Hero of the Soviet Union medals. Two pilots, Katya Ryabova and Nadya Popova, carried out eighteen raids in one night. In total, the Night Witches flew more than 24,000 missions. Most of the women pilots who survived the war flew hundreds of missions.

SEE ALSO:
- Aerobatics • Biplane • Bomber
- Pilot • World War II

Ornithopter

An ornithopter is a machine with flapping wings. Early inventors tried to copy bird flight by designing and building these aircraft, but their designs failed to get off the ground. Small ornithopters have flown as toy models and research experiments.

The principle behind the ornithopter is that the flapping wings provide both lift and propulsion. When people first dreamed of flying, they naturally tried to imitate birds, and some tried strapping wings to their arms. In about 1500, Italian artist and inventor, Leonardo da Vinci, made a sketch of a practical-looking ornithopter, but it never flew. One of the first toy airplanes was a model flown at the court of the King of Poland in 1647, by an Italian inventor named Titus Burratini. His model apparently had fixed and flapping wings.

Flapping wings were not the answer for human flight, as glider pioneer Sir George Cayley (1773–1857) realized. Cayley decided that if an aircraft needed wings for lift, some other means must be found for propulsion. The answer was the fixed-wing airplane with a propeller. Cayley never flew a powered plane, however, and people continued to design flapping-wing machines.

⟲ In 2006, an ornithopter designed by James Delaurier made a short, sustained flight.

A model ornithopter, flown in 1870 by Gustave Trouvé (1839–1902), was powered by revolver cartridges. The exploding cartridge forced the wings down, and springs pushed the wings up again. Trouvé's ornithopter apparently flew for 230 feet (70 meters). Another French inventor, Alphonse Pénaud (1850–1880), designed an ornithopter as well as model gliders. His models flew well and later inspired Orville and Wilbur Wright. Unfortunately, when his clever designs for flying machines were not taken up by the authorities, Pénaud became depressed and committed suicide. In the 1890s, Australian Lawrence Hargrave built an ornithopter powered by using steam or compressed air to flap one set of small wings, while relying on large, fixed wings for lift.

A toy ornithopter flies because it is so lightweight. By the 1930s, rubber-powered model "flapping birds" were popular toys. These ornithopters could fly well in calm air for a short period of time. A small ornithopter, made from wood, wire, paper, or plastic, needs only a taut, twisted rubber band for power to make the wings flap. Model ornithopters fly best in the calm air inside a large building—rubber-powered ornithopters have achieved inside flight durations of more than 20 minutes. For outside use, there are radio-controlled ornithopter kits that are usually powered by a small electric motor.

The problem with ornithopters is finding a power plant that will make the wings beat up and down with sufficient power to generate both lift and propulsion but is not so heavy that the machine cannot leave the ground. A leading designer of ornithopters is Dr. James Delaurier of the University of Toronto's Institute for Aerospace Studies in Canada. His team has flown several small flapping-wing designs. In 2006 they successfully flew a larger machine, using a jet-assisted takeoff.

TECH TALK

HOW AN ORNITHOPTER FLIES

An ornithopter creates all its thrust—and most of its lift—by flapping its wings. The wings beat in a twisting motion rather than directly up and down. They are joined by a center section that is moved up and down by the drive mechanism from the engine. The thrust comes from a low-pressure zone around the leading edge of the wing that generates leading-edge suction. Many toy ornithopters fly nose-up to ensure enough lift. The tail is usually set at a steep angle of incidence, angled up.

SEE ALSO:
• Aerodynamics • Bird • Cayley, George • Da Vinci, Leonardo • Wing

Parachute

A parachute is a canopy that slows the fall of an object or person through the air. The word *parachute* means "against a fall."

Parachutes have saved the lives of many pilots who needed to eject from damaged airplanes. Parachutes are used to drop supplies and paratroopers (parachuting soldiers) from airplanes. Sports parachutists enjoy freefall skydiving. Yet another use for a parachute is as an airbrake, to slow an airplane, spacecraft, or other vehicle as it lands.

A personal parachute is packed in a bag or body pack worn by the parachutist and attached to a strong harness or supporting rig. After exiting the aircraft, the parachutist opens the parachute by pulling a handle called the ripcord. Parachutes also can be opened automatically. When a pilot ejects from a jet plane, for example, the ejector seat mechanism opens the parachute. Brake parachutes for slowing down an airplane are stowed in the tail and open only after the plane has touched down on the runway. A spacecraft parachute opens after reentry into the atmosphere. Other brake parachutes may be automatic or manually deployed.

The canopy is made of a tough, light fabric—silk was traditional, but nylon and other synthetic materials are used today. The traditional shape for a parachute canopy was a circle, but modern parachutes are usually square or rectangular. The parachutist's harness is attached by straps, called risers, to suspension lines around the edge of the canopy. As parachutists float to the ground, they can make turns by tugging on steering lines.

The Invention of the Parachute

Even a handkerchief will act as a parachute if strings are tied from each corner to a small weight. The idea may have struck someone far back in history. The Chinese invented the umbrella, and they may have adapted the umbrella shape to try parachuting 1,000 years ago. In around 1485, Italian artist and inventor Leonardo da Vinci drew a cone-shaped

◗ A drawing from the early 1800s shows three views of André Jacques Garnerin's 1797 parachute: the top, the release from a balloon, and the parachute floating down after release.

parachute, but it is not known whether his device was ever tested.

The first parachute jumps recorded in Europe were made in 1783 by Sebastien Lenormand of France, who dropped weights and animals from a tower using a parachute that looked rather like a lampshade. In 1797, André Jacques Garnerin made a circular parachute of cotton cloth, from which hung a basket for a passenger. On October 22, 1797, he and his parachute were carried aloft by a balloon. Garnerin descended safely from about 2,000 feet (610 meters) above the city of Paris.

In the nineteenth century, parachute jumping from balloons became a popular form of entertainment. Balloon jumper Charles Broadwick invented the first body-pack parachute in 1905. The parachute pack was fastened to the balloon by a line. As the jumper fell, the line tightened and pulled the parachute canopy open. In 1912, Captain Albert Berry made the first parachute jump from an airplane, at a height of 1,500 feet (460 meters) above St. Louis, Missouri. Georgie Thompson, a teenager who jumped with Broadwick, was the first woman to jump from an airplane and land using a parachute, in 1913.

During World War I (1914–1918) few pilots had parachutes. Generals (and many pilots) argued that parachutes were too cumbersome. Military personnel who went up in observation balloons did have parachutes, however, so they could leap out if their balloons were hit by enemy gunfire.

HOW A PARACHUTE WORKS

When a parachute opens, air pushes up to fill the canopy. The air acts against the force of gravity and slows the fall of the object to which it is attached. A parachute increases air resistance because it offers a large surface area that produces friction with the air. At first, friction is greater than gravity, so the parachutist slows down. When the friction decreases to the point at which it is equal to the force of gravity, the parachutist descends at a constant speed. In certain weather conditions, the upward force of air may push the parachutist upward for a short time.

The Modern Parachute

In 1919, the forerunner of the modern parachute was tested in the United States by a group of jumpers, including James Floyd Smith, Leo Stevens, and Leslie Irvin. The new parachute had a circular canopy and a smaller parachute called a pilot chute. Both parachutes and their lines were folded and stowed in a cloth pack. The pack was held closed by three metal pins attached to a wire ripcord. When the jumper tugged a handle on the harness, the ripcord ripped the pins free, and the pack opened. The pilot chute flew out, acted as a brake, and pulled out the main canopy. On April 28, 1919, Leslie Irvin tested the parachute after jumping out of a plane over McCook Field in Ohio. In 1922, came the first use of a parachute in an emergency when an American military pilot, Lieutenant Harold Harris, bailed out of a test plane over North Dayton, Ohio. Throughout the 1920s, barnstormers and show jumpers made parachute jumps to entertain crowds at flying shows.

Most jumps were from low level. Doctors warned that parachuting from great heights, or falling at high speed before the parachute opened, would kill the jumper. In fact, such fears were proved wrong. In 1945, Lieutenant Colonel William Lovelace jumped from a B-17 bomber at a height of 40,000 feet (12,190 meters). Although he wore breathing apparatus, Lovelace became unconscious, but his parachute opened, and he landed safely.

Paratroops and Ejector Seats

The military began to realize the tactical importance of parachutes for landing both troops and supplies from aircraft. Airborne units were formed and used in World War II (1939–1945). The Germans used paratroops to attack Crete in 1941. In 1944, thousands of Allied airborne troops were dropped from the skies above Europe during the D-Day and Arnhem assaults. Transport planes also parachuted supplies to soldiers and dropped food and medicines to civilians.

⮌ Since World War II, parachutes have been used by the military to get troops and supplies into difficult places. This photograph shows a team of U.S. and Canadian pararescuers using parafoils during a search-and-rescue exercise.

⮡ The 150-foot (46-meter) solid rocket boosters used to launch the Space Shuttle are retrieved for reuse after they travel back to Earth with the help of parachutes.

During World War II, Allied fighter pilots and bomber crews used parachutes. Hundreds of combat fliers parachuted from planes, often after their planes had been hit by enemy fire. After the U.S. Doolittle Raid on Tokyo in 1942, all but one of the B-25 crews taking part had to use parachutes when their planes ran out of fuel over China.

In the 1940s, new parachute techniques were invented for jet pilots. The ejection seat, first tried in 1946, was available by 1951 with a pressure-operated parachute that would open at a preset, safe altitude. All a pilot had to do was jettison the cockpit cover and pull down a face blind; he and his seat were ejected from the airplane, and the parachute opened. Ejection seats are now standard in most military airplanes.

Other Kinds of Parachutes

Many jet planes have very high landing speeds, so tail parachutes are used as extra brakes. Spacecraft returning to Earth have used parachutes to break their fall. For the astronauts of the 1969–1972 Apollo missions, the last stage of their journey was the slowest. They dangled in a capsule beneath billowing landing parachutes that dropped the spacecraft into the ocean, thereby adding the word "splashdown" to the nation's vocabulary.

The modern wing-parachute or parafoil is highly maneuverable, and parachute jumping has now become an international, competitive sport for individual jumpers and teams. Freefall skydiving is a thrilling spectacle at air shows and also has become an enjoyable recreation for many enthusiasts.

SEE ALSO:
• Barnstorming • Ejection Seat
• Skydiving • Takeoff and Landing

Pilot

A pilot is the person who controls an aircraft. The name "pilot" was originally given to someone who steers a ship. Pilots fly everything from large airliners and fast military jets to airships and balloons. They also fly light aircraft, business jets, cargo planes, crop dusters, search-and-rescue helicopters, air ambulances, and other aircraft types. Requirements for a good pilot are sharp eyesight, intelligence, and calm judgment. All pilots must be physically fit and mentally alert.

Pioneers and Celebrities

The first pilots taught themselves to fly and were often their own mechanics as well. The world's first aero club was set up in France in 1898—five years before the Wright brothers' famous 1903 flight. The Aero Club of America was founded in November 1905. Among early aviators were Glenn Curtiss—who flew his June Bug airplane for the first time in 1908—and Louis Blériot, the first airplane pilot to fly from France to England, in 1909. The world's first international aviation meet, at Reims, France, in the summer of 1909, saw just twenty-three airplanes flying.

During World War I (1914–1918) airplane pilots earned a reputation for gallantry and chivalry. Fighter "aces," such as American Captain Edward V. Rickenbacker, dueled in the skies. The first African American combat pilot, Eugène Bullard, was denied entry into the U.S. Army Air Corps on racial grounds and flew instead with the French Flying Corps. After the war, barnstormers (stunt pilots) thrilled crowds across the country with aerobatic shows. Women also took to the air. Ruth Law, an American pilot, was the first woman to loop the loop, in 1916. Bessie Coleman was the first female African American pilot.

⤴ Eugène Jacques Bullard, the first African American combat pilot, flew for France during World War I. He later fought with the French Resistance in World War II and returned to the United States after being wounded.

EARLY WOMEN PILOTS

The world's first ever female pilot was Elise Raymonde Delaroche of France, who received Pilot's Certificate Number 36 in March 1910. She was killed in an airplane accident in 1919. Harriet Quimby, the first female American pilot, gained her pilot's license in August 1911. On April 16, 1912, Quimby became the first woman to fly an airplane across the English Channel between England and France.

⬇ Harriet Quimby sits in the cockpit of an airplane in 1911.

Men and women pilots flew stunts for movies—racing trains and flying under river bridges—and competed in air races. Record-breaking flights turned pilots into celebrities. Charles Lindbergh (the first pilot to fly solo across the Atlantic Ocean, in 1927) and Amelia Earhart (the first woman to achieve this feat, in 1932) were as famous as movie stars. Ruth Nichols was the first woman pilot to land in every one of the states of the United States. Clyde Pangborn and Hugh Herndon flew nonstop across the Pacific Ocean in 1931, and Wiley Post circled the world solo in 1933.

⮕ In 1929, James H. Doolittle (1896–1993) made the world's first instruments-only takeoff, level flight, and landing. In 1932, he set a world speed record for land planes. During World War II, he led the first bombing raid on Tokyo, Japan, and later commanded the Eighth Air Force in Europe and on the island of Okinawa, Japan.

THE TUSKEGEE AIRMEN

Until 1940, African Americans were not allowed to fly in the U.S. military. In 1941, however, the U.S. Army Air Corps formed an all-black unit in Tuskegee, Alabama. Ground crew, navigators, pilots, and weapons crews were rigorously trained for combat at the Tuskegee Army Air Field (TAAF) and elsewhere in the United States. By 1946, almost 1,000 pilots had completed training. About 450 of these men served overseas during World War II. The Tuskegee airmen, as they became known, achieved an outstanding record, gaining respect in an era when prejudice, segregation, and lack of opportunity were the norm for African Americans. They flew thousands of missions, destroyed over 1,000 enemy aircraft, and received hundreds of medals, including more then 150 Distinguished Flying Crosses.

This group of Tuskegee airmen were pilots with the 332nd Fighter Group stationed in Italy during World War II.

The Growth of Aviation

As the airline industry grew in the 1920s and 1930s, barnstormers and air racers often became commercial pilots. Some of them entered the military. During World War II (1939–1945), many military pilots learned to fly straight out of college and were often pitched into combat after only a few weeks of training. Fighter pilots in particular earned hero status. Most combat pilots were men, while female pilots delivered airplanes from factories and transported soldiers. After the war, test pilots broke new ground flying the jet- and rocket-powered planes of the supersonic era. Most of the first astronauts selected in the 1960s for the U.S. space program were ex-test pilots.

By the 1970s, with air traffic growing rapidly, the job of the commercial pilot became more demanding. Men still dominated the cockpit, but a few women started flying airliners. Ruth Nichols flew commercial planes as early as 1932. The first regular woman pilot for a U.S. scheduled airline was Emily Warner, who piloted Boeing 737s for Frontier Airlines in 1973. By the end of the twentieth century, military forces had women stationed alongside men in combat. The first American woman pilot to drop bombs in combat was Lieutenant Kendra Williams of the U.S. Navy, during Operation Desert Fox in Iraq (1998).

Learning to Fly

Many flying students start with a short introductory lesson at a flying school. No pilot certificate or medical certificate is needed for a trial flight, but these are required if a student continues and wants to fly solo. To be certified fit to fly, a student must consult an aviation medical examiner approved by the Federal Aviation Administration (FAA). There are three classes of medical certificate: class 1 for airline pilots, class 2 for other commercial pilots (anyone paid to fly), and class 3 for recreational pilots.

At first, the beginner student flies with an instructor in a two-seat plane, but does most of the actual handling of controls. The trainee pilot must obtain a student pilot certificate, issued by the FAA. Only sports pilots (flying microlights or similar airplanes) in the United States can fly on the basis of a motor vehicle driver license. The student pilot must pass a written test and learn to perform certain maneuvers—including takeoff and landing—before being allowed to fly solo. To gain a private pilot certificate, a person must be at least seventeen years old.

Pilots in the United States may not carry passengers unless they have a recreational pilot certificate or a private pilot certificate. Obtaining these certificates can cost several thousand dollars. Flight training is usually charged by the hour, and most students need 40 to 60 hours for private pilot training. It takes less time to obtain a recreational pilot certificate, but this restricts pilots in certain ways (for example, they cannot fly where communications with air traffic control are required).

○ Pilots learn to fly with an experienced instructor, who also has a set of controls. Student pilots in the United States must be at least sixteen years old to fly solo.

Students also must pass a written exam on a computer. The FAA provides information needed to gain a certificate. Study materials include information on weather, airplane flying, glider flying, balloon flying, and rotorcraft flying. The final flight exam, or check ride, is done with an examiner and includes a question-and-answer session and a flight test lasting up to 1½ hours.

Becoming Professional

Professional pilots must have an instrument rating. For this, they need to do at least 50 hours of cross-country flying (between one airfield and another). They must be able to fly by visual flight rules (VFR) and instrument flight rules (IFR) using electronic aids. IFR involves instrument training, during which pilots fly using just their instruments so that they will be able to fly even when visibility is poor or zero. Commercial pilots also familiarize themselves with the use

of radio, radar, and the landing systems used at airports, such as the microwave landing system (MLS) and the older instrument landing system (ILS).

Before gaining a commercial pilot license (CPL), a pilot must have completed at least 250 hours of flight time, recorded in a personal log book, and must have learned to fly more complex aircraft (with flaps and retractable landing gear). The flight examination includes two flight sessions: one in a training aircraft and another in a more complex airplane, although a student may fly the entire test in the complex airplane. Many commercial pilots add a multi-engine rating, which they need to fly aircraft with more than one engine.

Airline Pilots

Every airplane is slightly different to fly, so pilots have to qualify in every kind of plane they have not flown previously. Initial training for pilots joining an airline takes about ten weeks, during which they learn the specific procedures of the airline and get used to the aircraft. Training is often done in pairs and includes simulator training and practice in all maneuvers. After pilots pass this training course, they receive their initial operating experience in the air alongside an instructor pilot. They take a final flight test or "line check" and are then cleared to fly scheduled passenger flights.

Statistically, flying is very safe. According to the FAA, a well-built, well-maintained aircraft flown by a competent and prudent pilot is as safe as any other form of transportation.

FLIGHT SIMULATORS

Before 1930, pilot training consisted of ground instruction followed by flights in a dual-control airplane with an instructor. To improve training, the flight simulator was invented. The first was the 1930 Link Trainer, a mechanical simulator that gave a trainee pilot the feel of airplane motion. This simulator was improved by the addition of instrument simulation. The celestial navigation trainer (1941) showed bomber crews how to fly at night. In 1948, Pan American pilots learning to fly the Stratocruiser airliner trained in a cockpit replica with a full set of instruments. From this were developed full motion simulators, which gave the trainee a picture of the ground while practicing approaches to the runway and other maneuvers. By the 1970s, simulators with hydraulic actuators could control each axis of motion, so the trainee pilot experienced a full range of airplane motions, including roll, pitch, and yaw. Computers and electronic display technology can now create a realistic virtual skyscape and landscape. Simulators are useful for training flight crews in operating procedures and for exposing pilots to risky situations, such as a complete engine failure, which cannot be practiced in a real airplane.

⟳ Flight simulators have a full range of controls and a view of a virtual-reality world outside.

STEVE FOSSETT'S RECORD FLIGHTS

American pilot Steve Fossett set remarkable records piloting balloons and specialized airplanes. In 1995, he made the first solo flight across the Pacific Ocean in a balloon. In 2002, he made the first solo, nonstop, round-the-world balloon flight (in 14 days and 19 hours). In 2005, Fossett piloted the Virgin Atlantic *Globalflyer* on the first nonstop, solo, round-the-world airplane flight, a trip that took 67 hours and covered 22,878 miles (36,811 kilometers). In February 2006 Fossett then set a record for the world's longest flight, when he flew *Globalflyer* for 26,389 miles (42,460 kilometers) in a journey lasting nearly 77 hours. In September 2007, Fossett disappeared in a small airplane while in a scouting flight over the Nevada desert. He was officially pronounced dead in February 2008.

A modern aircraft is a highly complex, computerized machine. To fly it properly, a pilot needs technical as well as piloting skills. A first officer and other cabin crew assist the captain of an airliner. Most large commercial airplanes have two pilots. (General aviation airplanes and helicopters are usually flown by a single pilot.)

Heavy Responsibilities

The job of an airline pilot can seem exciting. Pilots jet around the world, and they are well paid, but the routine involves hard work, a lot of waiting time, and heavy responsibilities. The pilot and first officer's tasks include figuring out a flight plan showing the route, flying height, and fuel capacity. They supervise loading and fueling of the aircraft, brief the cabin crew, and carry out preflight checks. Airline pilots must communicate constantly: with air traffic control before takeoff, during the flight, and while landing, and with their passengers during the flight. They check the aircraft's technical performance, and position, the weather, and air traffic. At the end of a flight, pilots update the aircraft logbook and write reports about any incidents during a flight.

At all times, an airline pilot must be ready to act promptly should an emergency occur. A pilot is responsible for the safety of the aircraft and its passengers. In the wake of the terrorist attacks in New York City in September 2001, airspace security was tightened up to protect potential terrorist targets.

⮫ Most airline pilots in the United States belong to the Air Line Pilots Association, a labor union and professional organization for pilots founded in 1931.

Pilots are aware that the FAA may impose temporary flight restrictions (TFRs) to restrict aircraft movements in certain areas, for example around air shows, space launches, forest fires, or presidential visits. TFRs also protect potential targets, such as military bases and government installations.

Military Pilots

Military pilots fly with the U.S. Air Force, U.S. Navy, U.S. Marine Corps, U.S. Army, U.S. Coast Guard, and the Air National Guard. In the U.S. Air Force, the Air Education and Training Command (AETC) is based at Randolph Air Force Base near San Antonio, Texas. Personnel hoping to become pilots receive up to 25 hours of initial flight training from civilian instructors. Selected candidates move on to further training by military instructors. Student pilots learn to fly on fairly slow training airplanes, such as the turboprop T-6II, moving up to the twin-jet T-37, and then to a supersonic jet such as the T-38.

All students learn basic flight skills. Then they are selected for one of several advanced training tracks, depending on the type of aircraft they will fly. Helicopter pilots, for example, receive special training, on the UH-1 Huey helicopter. Student airlift (transport) and tanker pilots train on the T-1A Jayhawk, and others fly the T-44 to learn how to pilot a multi-engine, turboprop airplane such as the C-130 Hercules.

Pilots complete their training at U.S. Air Force bases around the country. For example, fighter pilots qualifying from the T-38 course at Randolph Air Force Base go on to fly the F-15 Eagle at Tyndall Air Force Base, Florida, or the F-16 Fighting Falcon at Luke Air Force Base, Arizona. On completion of their military service, many pilots continue to fly as civilian pilots.

SEE ALSO:
- Barnstorming • Blériot, Louis
- Coleman, Bessie • Curtiss, Glenn
- Earhart, Amelia • Lindbergh, Charles • Night Witches

Pitch, Roll, and Yaw

Pitch, roll, and yaw are the three ways in which aircraft and spacecraft can change their direction of flight. Pitch, roll, and yaw are rotations. When something rotates, it turns around an imaginary line called an axis.

Defining Pitch, Roll, and Yaw

Imagine an airplane with a stick pushed through it from nose to tail, so the plane can spin on the stick. This type of rotation is called roll, and the stick is the roll axis. A stick pushed through a plane from wingtip to wingtip is the pitch axis. A stick pushed through a plane from top to bottom is the yaw axis. The position, or angle, of an aircraft—the amount of pitch, roll, and yaw it has—is called its attitude. In a standard airplane, the pilot generally can change the plane's attitude by operating controls that move the elevators, ailerons, and rudder.

Changing an airplane's pitch makes its nose tip up or down. The pilot changes the pitch by using the control stick to tilt the elevators, which are at the back of the tail. Pulling the stick back tilts the elevators up. Air flowing over them pushes the plane's tail down and raises its nose. Pushing the stick forward has the opposite effect.

Rolling, or banking an aircraft to one side, enables it to turn. When it rolls, the lift produced by the wings tilts to one side instead of acting straight upward. The sideways part of the lift pulls the plane into a turn. A pilot makes a plane roll by pushing the control stick to one side or (on a larger airplane) turning the yoke on top of the control column. This control moves the ailerons, which are at the back of the wings. When the

These diagrams show an airplane's three principal axes and how the plane rotates around them in pitch, roll, and yaw.

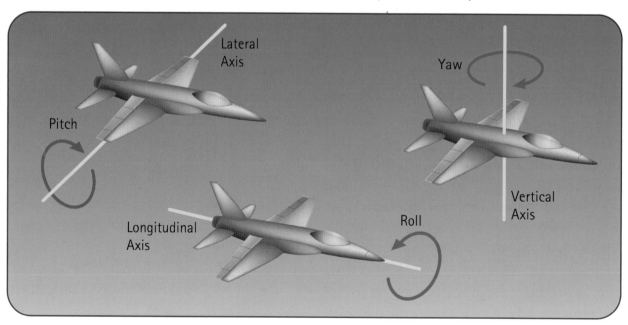

Lateral Axis

Pitch

Yaw

Longitudinal Axis

Roll

Vertical Axis

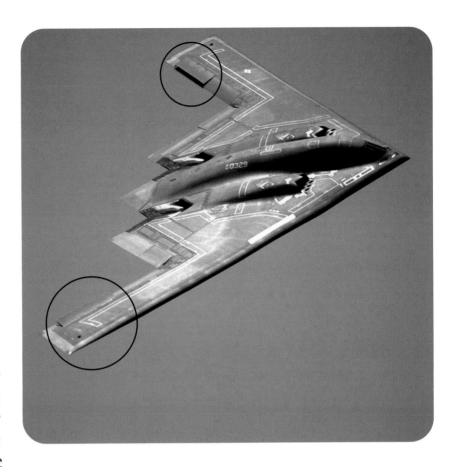

⮕ The B-2 Spirit Bomber (along with other flying wing aircraft) has elevons (circled) at the back of its wings that function as combined ailerons and elevators.

aileron on one wing tilts down, the wing rises. The aileron on the other wing tilts up, and the wing sinks.

Yaw is the name for the motion when a plane's nose turns to the left or right. Yaw is controlled by a rudder, which is mounted on an airplane's tailfin. The rudder is operated by a pair of foot pedals. Turning the rudder to one side pushes the plane's tail to the opposite side and swings the nose around. When a plane rolls into a turn, the pilot also adjusts its pitch and yaw to keep the plane's nose pointing in the right direction.

Alternative Control Surfaces

Some aircraft do not have ailerons or elevators because they do not have normal wings and tail. Flying wing aircraft have no tail at all. These aircraft use a different type of control surface called an elevon, which is a tilting part at the back of the wing. Elevons combine the jobs of the ailerons and the elevators. If the elevons tilt up or down together, they work like elevators to raise or lower a plane's nose. If they work in opposite directions, one up and one down, they work like ailerons and make a plane roll.

Most airplanes have two small wing-like parts, one on each side of the tailfin. They are called tailplanes, or horizontal stabilizers. An elevator at the back of each stabilizer tilts up or down to control pitch. Some planes have a different type of tailplane. The whole tailplane tilts instead of just the elevator. It does the job of the stabilizer and elevator together and is therefore sometimes called a stabilator. Other names for this part are all-moving tailplane or all-flying tailplane.

In some fighter planes, the pitch is controlled by small, tilting winglets on the nose known as canards.

Trimming

The weight of a whole airplane acts as if all of its mass were concentrated at one point. This point is called the center of gravity. If an airplane could hang from a wire at this point, it would hang perfectly level. A flying airplane is held up by lift. All the lift acts at one point called the center of pressure. Airplanes are designed so that the center of pressure and center of gravity are close to each other, but it is not possible to get them in exactly the same place. Both of them move during a flight as a plane's speed and attitude change and as it uses up fuel. Passengers walking about in an airliner move its center of gravity as well.

If the center of pressure is in front of the center of gravity, it pulls the plane's nose up and the plane is said to be tail-heavy. If the center of pressure is behind the center of gravity, it pulls the plane's nose down. Then the plane becomes nose-heavy. If a plane were consistently nose-heavy or tail-heavy, the pilot would have to keep pulling or pushing the control stick for long periods to stay level. To avoid this, planes have trim control. This is usually a wheel, lever, or group of switches in the cockpit. Pilots use trim control to move the center of pressure forward or backward until it lines up with the center of gravity, making the aircraft level. Small airplanes have panels in the tail, called trim tabs. They tilt up or down to move the tail up or down until the plane is level. Larger airliners trim their pitch by tilting the horizontal stabilizers in their tail. A level plane is called a trimmed plane.

Helicopters and Spacecraft

Helicopters also change their attitude by pitch, roll, and yaw motions, but they do it differently from fixed-wing airplanes. Unlike airplanes, helicopters do not have wings, ailerons, elevators, or a rudder. Instead, they use their main rotor, the spinning blades on top, for pitch and roll movements. Tilting the whole rotor forward or backward changes the pitch of a helicopter. Tilting the rotor to one side or the other makes the air-

⤺ Helicopters use their main rotor, or spinning blades, to change their pitch and roll. A smaller rotor in the tail alters yaw.

POINTING AT THE HORIZON

During daylight, pilots often keep an aircraft's attitude under control by simply looking out of the window. To keep an airplane flying straight and level, the pilot keeps its nose pointing at the horizon and the wings level with the horizon. This is called visual flight. If the horizon is not visible because of darkness or clouds, an instrument in the cockpit called the artificial horizon is used instead. It shows an outline of the plane's wings in front of a ball with the horizon marked on it. Whatever way the plane moves, the ball rotates to keep its artificial horizon in line with the real horizon. A glance at this instrument shows whether or not the plane is level.

craft roll. Speeding up or slowing down the small rotor in the helicopter's tail makes the aircraft yaw to the left or right.

The Space Shuttle looks like a delta-wing airplane. It uses elevons in its wings to control pitch and roll. A rudder in its tailfin controls yaw. It also has a hinged panel called the body flap under its tail. No other aircraft or spacecraft has this control surface, which is used to trim the Space Shuttle's pitch. In space, the Space Shuttle is unable to change its attitude in the same way as a plane

🔽 The Space Shuttle's body flap is installed under the main engines. It provides the spacecraft with pitch control trim during its descent to Earth.

because elevons and rudders do not work in space. Instead, it fires small rocket thrusters in its nose and tail.

SEE ALSO:
• Aileron and Rudder • Control System • Lift and Drag • Tail

Pollution

Pollution is the process of making the environment dirty, dangerous, or in other ways unpleasant or unhealthy for people, animals, and plants. Flying contributes to pollution through the emissions from airplane engines and through noise and environmental damage around airports.

Transportation is a major source of air pollution in the United States and other industrial nations. Jet engines, like automobile engines, burn carbon-based fuel. During the burning process or combustion, airplane engines give off carbon monoxide, carbon dioxide, hydrocarbons (compounds of carbon and hydrogen), and nitrogen oxides (compounds of nitrogen and oxygen). These substances are all pollutants, and too many of them in the atmosphere can have damaging effects on people, on animals, on plants, and even on buildings. Polluted air is unhealthy to breathe. Heavy concentration of pollutants around cities can form smog, reducing visibility and air quality and endangering the health of people.

The white contrails of jet planes in the blue sky may not look as dirty as the smoky exhaust from a truck on the highway, but they contain harmful, polluting gases.

Scientists believe that carbon-based pollutants are causing damage to Earth's atmosphere. A buildup of carbon dioxide gas in the atmosphere from burning fossil fuels—such as gasoline and aviation fuel—is thought by many experts to contribute to the greenhouse effect. The gases trap heat from sunlight, therefore contributing to global warming and climate change.

High-flying jet aircraft emit those gases close to Earth's surface and at higher altitudes. The primary gas in jet engine emissions is carbon dioxide, which can linger in the atmosphere for up to a hundred years. Aviation emissions account for up to 4 percent of all global carbon dioxide emissions from the burning of fossil fuels. Carbon dioxide combined with other airplane exhaust gases could be having a much greater impact on the air than carbon dioxide alone.

Most legislation passed in recent decades to cut air pollution has been directed at industry and automobiles. With aviation growth at around 5 percent a year, however, the development of cleaner aircraft engines is vital.

Airports also are a source of pollution—not simply because of the number of airplanes using them, but because a busy airport draws in thousands of cars and trucks every day. Airport buildings and handling facilities consume a lot of energy and produce a lot of waste. Even the chemicals used to de-ice airplanes in winter pose a pollution risk to the soil and the water cycle.

NOISE POLLUTION

As air traffic increases, there are concerns about noise pollution. Anyone who has stood on a runway close to a jet plane taking off knows that it is very noisy. The loudness is measured in decibels. A jet plane taking off can reach 130 decibels. Supersonic planes also make a sonic boom. Protests about the boom ended airline plans to fly the Concorde on transcontinental supersonic flights in the 1970s. Modern turbofan engines are more efficient and less noisy than the engines of fifty years ago, but many airports suspend flights at night so that local residents can sleep undisturbed.

Some campaigners argue for cuts in flights or at least increased airport and airline taxes—and thus higher fares—to reflect the true environmental cost of flying. Aircraft manufacturers respond that new airplane engines are becoming increasingly efficient and clean. They also say the introduction of larger airplanes means fewer flights, less fuel burned, and therefore less pollution.

SEE ALSO:
• Aircraft Design • Airport • Engine
• Fuel • Future of Aviation

Pressure

Pressure is the pressing effect of a force acting on a surface. Scientifically expressed, pressure is the force per unit area acting on a surface. Pressure (P) is defined as the force applied (F) divided by the area (A) of application. The equation for pressure is: $P=F/A$.

When a force acts on a material (solid, liquid, or gas), the result is pressure. The causes of pressure are as varied as the causes of forces. The force of a party balloon squeezing the air inside it produces pressure. The weight of a book pressing down onto a table produces pressure. Oil forced through the pipes of an aircraft's hydraulic system produces pressure. Gravity pulling air against Earth's surface produces atmospheric pressure.

Atmospheric Pressure

Atmospheric pressure, or air pressure, can be measured in various ways. The weight of air pressing down on the Earth's surface produces an air pressure at sea level of 14.7 pounds per square inch (psi), or about 100 kilopascals (100,000 pascals). Meteorologists (weather scientists) measure pressure in bars. The air pressure at sea level is about 1 bar, or 1,000 millibars. This pressure also is known as "1 atmosphere."

Air pressure in the atmosphere falls with increasing height. Gravity pulls air against Earth's surface. Air at Earth's surface has the weight of all the rest of

THE BAROMETER

Atmospheric pressure is measured with an instrument called a barometer. The first barometer was made in 1643 by an Italian scientist named Evangelista Torricelli (1608–1647). He filled a long glass tube with mercury. Then he turned the tube upside down with its open mouth in a bowl of mercury. Some of the mercury ran down into the bowl, but not all of it. A column of mercury about 30 inches (76 centimeters) high stayed in the tube. Its weight was balanced by air pressure acting on the mercury in the bowl. Torricelli realized that changes in the column's level were due to changes in atmospheric pressure. Mercury barometers work in this way.

An aneroid barometer works in a different way. It is a sealed can with some air taken out. Atmospheric pressure squashes the can. The amount of squashing changes when the air pressure changes. These small movements are linked to a needle pointing at a pressure scale. Because they do not need a tall tube of mercury, aneroid barometers are much smaller than mercury barometers.

A Learjet flying at 41,000 feet (12,500 meters) must be pressurized. At that altitude, passengers would lose consciousness without pressurized air.

the air above it bearing down on it, so the pressure is greatest here. Air higher in the atmosphere has less air from above pressing down on it, so the air pressure higher above the ground is lower. This is an important factor to consider for a person flying high in the atmosphere or going into space.

One-fifth of air, or about 20 percent, is oxygen. The thin, low-pressure air at the top of a high mountain contains the same percentage of oxygen as air near the ground, but because there is less air at high altitude, there is also less oxygen. The human body is very sensitive to sudden, even small, changes in pressure. Going up a tall building in a fast elevator can make someone's ears pop. The shortage of oxygen in low-pressure air at high altitudes can cause more severe effects. When people go higher in the atmosphere, they may experience a variety of problems due to low air pressure.

Mountain climbers can suffer headaches, nausea, and dizziness when at altitude.

Adjusting Pressure

Big drops in air pressure are even more serious. Pilots and passengers of high-flying aircraft need protection from the low air pressure outside of the plane. Early airliners did not fly higher than about 10,000 feet (3,050 meters)—above that height, some passengers began to feel faint. To fly higher safely, an aircraft has to be pressurized. Extra air is pumped inside a modern airliner to raise the pressure. The air inside a pressurized airliner is not at sea level pressure. It is the same as the pressure at an altitude of about 8,000 feet (2,500 meters), which is about 11 psi, or 75.8 kilopascals.

Fighter pilots sit in a pressurized cockpit, but they also wear an oxygen mask in case the canopy shatters and the cockpit loses pressure.

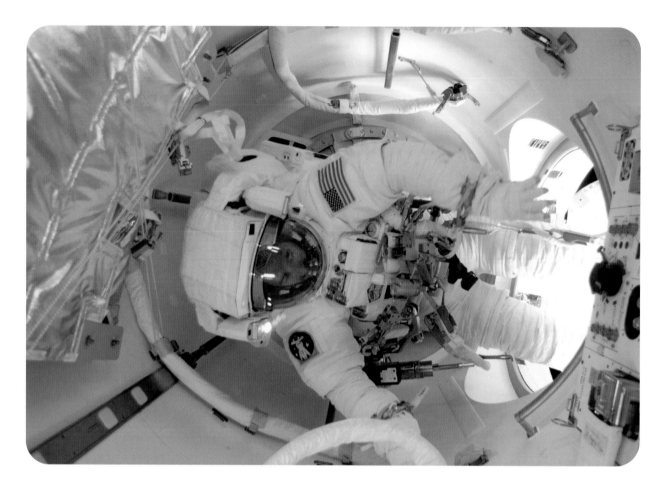

If an airliner suffers a sudden loss of pressure at its cruising altitude of about 35,000 feet (10,600 meters), oxygen masks drop down automatically from the ceiling. The crew and passengers have about 30 seconds to put them on before losing consciousness.

The Space Shuttle and International Space Station (ISS) are both pressurized to sea level pressure. The spacesuits worn by astronauts, however, are pressurized to only 4.3 psi (about 30 kilopascals) to prevent them from blowing up like a balloon. If an astronaut breathed air at such a low pressure, there would not be enough oxygen, so the suit is supplied with pure oxygen.

An astronaut on the International Space Station in 2001 tests an airlock that allows the crew to leave the station on space walks.

Because the pressure in a spacesuit is much lower than the pressure inside the spacecraft, the air pressure around astronauts preparing to leave their spacecraft is lowered in stages so that they can adjust safely. The day before a spacewalk, the air pressure inside the Space Shuttle is lowered to 10.2 psi (70 kilopascals). Space Station astronauts spend several hours inside an airlock where the air pressure is even lower, to prepare for the lower pressure inside their spacesuits.

HIGH RISKS

Early balloonists were the first aviators to discover the hazards of high-altitude flight. On September 5, 1862, Henry Coxwell and James Glaisher made a balloon ascent to more than 30,000 feet (9,100 meters). No manned balloon had flown as high before. As they passed 29,000 feet (8,850 meters), Glaisher became paralyzed. Then he lost consciousness. Coxwell lost the use of his arms and had to use his teeth to pull the rope that released hydrogen from the balloon and let them descend. They both survived.

Mechanics change the engine hydraulics pump on a cargo plane. Cargo planes also use hydraulics to operate their cargo doors and loading ramps.

Using Pressure

Aircraft and spacecraft can use different kinds of pressure in their operations. When a large liquid-fuel rocket is fired, pumps feed fuel to the engine. Pumps are too big and heavy for small rockets and spacecraft, so they use pressure instead. A high-pressure gas, such as helium, pushes the fuel from its storage tank to the engine.

Aircraft use high-pressure oil in their hydraulic systems. Hydraulic systems are those operated by liquid. Liquids cannot be squashed as much as gases because their molecules are already close together. This causes liquids under pressure to transmit force from one place to another. A digging machine works in this way. Oil pumped through flexible hoses operates mechanical pushers called rams, which move the digger's arm. Modern automobile brakes also work by hydraulic power. An aircraft's hydraulic system uses oil pressure to move its control surfaces and raise its landing gear. Cargo planes use hydraulic pressure to operate their cargo doors and loading ramps. Rockets use hydraulic power, too, when they swivel their engine nozzles for steering.

SEE ALSO:
- Air and Atmosphere • Altitude
- Astronaut

Propeller

A propeller is a set of long blades attached to a hub in the center. The job of a propeller is to change the turning force, or torque, of an engine into thrust. Thrust is the force that moves an aircraft through the air.

Propellers and Engines

More than 100 years ago, the first airplanes were powered through the air by propellers. When the jet engine was invented, it looked like the propeller's days might be over. Even in the age of jets and rockets, however, propellers are still widely used. A piston engine driving a propeller is still the best way to power a small plane today.

There also are many turbine-powered planes with propellers. A turboprop engine runs a lot faster than the best speed for the propeller it controls, so the engine and propeller are connected by a gearbox. Just as a car's gearbox lets its engine and wheels run at different speeds, a turboprop's gearbox allows the engine to run at its ideal speed and the propeller to spin more slowly.

How Propellers Work

When a propeller spins, its blades cut through the air. The blades work like wings standing on end, whirling around in a circle and producing thrust in a forward direction.

The first airplane propellers were made from a solid block of wood. The angle, or twist, of these early propellers'

TWISTING BLADES

To work most effectively, propeller blades (like wings) have to meet the air at the right angle, which is called the angle of attack. The blades follow a complicated path through the air. They are rotating at the same time as the aircraft moves forward, so they follow a spiral path through the air. In order for the blades to meet the air at the right angle, they have to be twisted.

A propeller's blades are twisted more near the center than at the tip. Anyone who has ridden a merry-go-round or carousel knows that someone at the outside edge goes a lot faster than someone near the center. The reason is that the person on the outside edge has to go a lot farther than someone near the center to make one spin of the merry-go-round. The two journeys take the same time, so the person at the edge travels faster. It is the same for a propeller. The tip of a blade goes faster than the part closer to the center, so the tip does not have to be twisted as much to create the same thrust as the rest of the blade.

blades could not be changed. The angle of the blades also is called the pitch, so these propellers are called fixed-pitch propellers. They worked most effectively at one particular speed. They did not work as well when a plane was moving slower or faster than this ideal speed.

Propellers with variable pitch allowed the blades to be twisted a little more or a little less so that the propeller worked better at different speeds. At first, the angle had to be changed by hand, and there were only two or three settings to choose from. Modern variable-pitch propellers automatically change the angle of the blades to suit the plane's speed.

If an aircraft engine breaks down during a flight, air rushing through the propeller blades keeps the propeller spinning. A propeller that spins freely in this way causes drag. Extra drag on one side of a plane acts like a brake and makes the plane turn to that side. Some propellers are designed to prevent this from happening. They have blades with edges that can be turned toward the air to cut their drag. This is called feathering. Some planes can actually reverse their propeller blades so that they blow air forward instead of backward, a process called reverse thrust. It helps cargo planes, such as the military C-130 Hercules, stop a short distance after landing.

The amount of thrust a propeller produces depends on the amount of air it pushes backward and how much it speeds up the air. Making a propeller bigger or spinning it faster produces more thrust. Adding more blades to a

A turboprop is a turbine engine that turns a propeller.

propeller also produces more thrust. The most efficient propellers move a large amount of air and speed it up a little.

Propeller planes are unable to increase their speed indefinitely. When they reach a speed of about 520 miles per hour (840 kilometers per hour), their propellers stop working so well. The problem is that the tips of the blades are moving faster than the speed of sound. The way that air flows around the blades changes at the speed of sound, and so propellers are no longer very good at producing thrust. The fastest that most propeller planes can fly is about 450 miles per hour (720 kilometers per hour).

Pusher Propellers

Modern propeller planes have their propellers at the front, pulling the plane through the air. These are called tractor propellers. In the early days of aviation, it was just as common for planes to have their propellers at the back, pushing the aircraft. The first successful powered airplane, the Wright brothers' *Flyer*, had two pusher propellers behind the wings.

Pusher propellers were popular because they did not get in the way of the air flowing over the wings. Without an engine or propeller in the way, the pilot had a great view ahead. It also was easier to fit guns in the nose of a fighter

PROPROTORS

The strangest propellers are called proprotors. They have this name because they work like helicopter rotors part of the time and like propellers the rest of the time. Compared to regular propellers, proprotors are enormous. The propellers of a C-130 Hercules cargo plane, for example, are 13.5 feet (4 meters) across. The V-22 Osprey's proprotors are nearly three times as big: 38 feet (12 meters) across. When the Osprey is on the ground, its engines are tipped up, pointing straight up in the air. Its two propellers look like helicopter rotors, and they work in the same way—when they spin, they lift the Osprey straight up into the air. Then the engines swivel forward, and the rotors work like propellers.

↻ The Osprey's engines and propellers tilt up (top left) for vertical takeoff. The engines tilt down and the propellers face forward (bottom left) for level flight.

The propellers of a C-130 Hercules go into reverse to slow it down when landing.

plane that had its engine and propeller mounted in back.

Pusher propellers had drawbacks, however. They made it harder for pilots to escape from planes in the air without hitting a propeller. Also, the heavy weight of an engine at the back of a plane could be dangerous in a crash. It could move forward and crush the cockpit. The problem of fitting guns to fighter planes with propellers at the front was solved during World War I (1914–1918) by a device called an interrupter gear. It stopped a machine gun from firing when a propeller blade was in front of the gun barrel. The gun was synchronized to fire through the spaces between the blades. By the end of the war, nearly all fighter planes had their propellers at the front.

Push-Pull Aircraft

A few planes were built with one propeller at the front and another propeller at the back. They were called push-pull aircraft. The German Dornier Do-335 of the 1940s was a push-pull fighter. With one engine and propeller at the front and another at the back, it was very fast. The push-pull layout has been used in experimental aircraft, too. *Voyager*, the first plane to fly around the world without landing anywhere or refueling on the way, was a push-pull plane. It had an engine and propeller in its nose and another set in its tail. Its 25,000-mile (about 40,250-kilometer) flight took nine days in 1986.

SEE ALSO:
• Engine • Helicopter • Thrust

Radar

Radar is a system that uses radio waves to detect and locate objects and movement. It has become a vital tool for safety and other purposes in aviation and spaceflight.

Air traffic control systems use radar to monitor aircraft movements and guide pilots safely. They use two types of radar. Primary radar locates an aircraft. Secondary radar transmits a signal that is received by a transponder (transmitter responder) in the plane. The transponder responds by sending back information about the aircraft, such as its call sign and current altitude.

Airliners and other aircraft are equipped with their own weather radar. The nose of an airliner contains a small radar antenna that scans the sky ahead of the aircraft and detects storms. Then the crew can change course as needed.

How Radar Works

The basic principle of radar is very simple. It sends out radio waves and then picks up any waves that are reflected back. Most radar systems are more complex, however, and they can tell much more about an object than just the fact that it is there. They can show its location, bearing (direction), range (distance), velocity, and altitude.

A radar system has four main parts. A transmitter produces radar signals.

⟳ A C-12 airplane used by U.S. Customs and Border Protection has a belly-mounted radar system that gives it long-range radar capability.

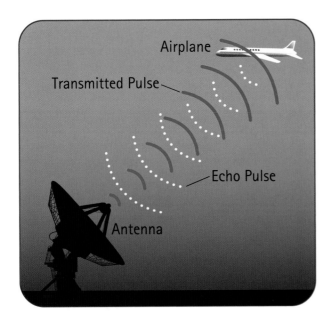

Airplane

Transmitted Pulse

Echo Pulse

Antenna

 A simple radar has an antenna that sends out signals in the form of pulses of radio waves. It picks up any echo pulses that come back and uses them to measure an object's distance and movement.

An antenna sends signals in the form of electromagnetic waves and picks up any reflections that return. A receiver amplifies the weak radar reflections and analyzes them. A display shows the received information on a screen.

Radar uses short radio waves called microwaves. The simplest type of radar is pulse radar. It sends out short bursts, or pulses, of radio waves and listens for any reflections that bounce back from a target, such as an aircraft. The direction from which the reflection comes shows the aircraft's bearing. The time the pulse takes to bounce back gives its range.

Antennae

A dish-shaped antenna can be steered to scan a particular area of the sky. It may swing back and forth, or it may rotate so that it scans the whole sky in all directions. The most modern radar systems use a flat antenna that stays fixed in one place. A flat antenna is constructed from

DOPPLER RADAR

When a police car races past sounding its siren, the sound rises in pitch as the car approaches and falls as it goes away. This is called the Doppler effect, and it happens with all kinds of waves, including microwaves. Radar equipment can be designed to make use of this effect. It can show if something is flying toward the radar antenna or away from it, and how fast. A type of radar called Doppler radar was developed in the 1960s. It uses continuous radar waves instead of pulses. Pulse-Doppler radar systems combine basic pulse radar systems with Doppler radar.

At first, Doppler radar was used mainly for weather forecasting. By the 1980s, Doppler weather radars were able to measure the speed and direction of raindrops inside clouds and storms. Portable Doppler radars carried on the back of trucks are used to study the most extreme weather systems, especially thunderstorms and tornadoes.

Dish antennae such as these at a tracking station in California swivel to pick up signals.

he suggested that this ability might be used to avoid collisions at sea, but there was no interest in his idea.

In 1922, the effect was rediscovered when a ship on the Potomac River in Washington, D.C., caused a disturbance to radio signals being sent across the river. An airplane was detected by radar for the first time in 1930.

As World War II approached, scientists in Britain and Germany stepped up their research into radar. The first practical radar system for air defense was developed in Britain by Sir Robert Watson-Watt in 1935. During the war, the British coastline was protected by a system called Chain Home. When the system detected approaching aircraft, the planes' positions were plotted on a map in a control room. Fighter pilots

Soldiers of the U.S. Army Signal Corps used this early radar system in Italy in 1944.

thousands of small, electronic transmit-and-receive modules, and the radar beam is steered electronically. These radars are called electronically scanned arrays, or phased arrays. They can scan far faster than a rotating dish antenna, they can track many more targets, and —with fewer moving parts—they are more reliable.

Advanced combat aircraft, such as the F-22, are equipped with electronically scanned array radar. They can locate and track multiple high-speed targets and pass on the target information to the aircraft's weapons systems.

The Development of Radar

The invention of radar can be traced back to experiments with radio waves carried out by physicist, Heinrich Hertz (1857–1894). Hertz discovered that radio waves passed through some materials and were reflected by others. In 1904, scientist Christian Hülsmeyer showed that radio waves could detect ships, and

were then given instructions by radio to guide them toward the incoming enemy aircraft. Germany also developed an air defense radar system, called Freya, during World War II. In addition, radar was also used to guide searchlights and anti-aircraft guns.

These early radar systems were too big and heavy to install in an aircraft. In 1939, however, scientists at Birmingham University, England, invented a device called a cavity magnetron, which enabled radar equipment to transmit and receive much shorter radio waves. This made it possible to build smaller and more powerful radar equipment, light enough to be carried by aircraft. As they developed, these systems had been known as RDF (radio direction finding). In 1942, the term *radar* (short for radio detection and ranging) came into use.

In 1943, British bombers were equipped with a radar system named H2S. It pointed downward and showed a map of the ground on a screen inside the aircraft. It enabled bombers to find their targets through the cloud cover. An improved U.S. system called H2X was developed in 1945.

Using Radar in Space

Since the early days of manned spaceflight, spacecraft have been able to dock (link up) with each other. The Gemini program carried out the first docking in 1966 as a step toward a successful Moon landing mission. Apollo Command Modules docked with the U.S. Skylab space station. The Space Shuttle, Soyuz

DEFEATING RADAR

During the Cold War, the United States and Soviet Union competed with each other to produce the most advanced military radar for their combat aircraft. They also developed ground-based radar to give early warning of a missile attack. This competition led to research into ways of defeating enemy radar. There are six main methods used to confuse or block radar systems:

• **Electronic jamming**: Sending out radio signals to block or swamp enemy radar.

• **Generating false targets**: Sending out radio signals that make extra, confusing information appear on enemy radar screens.

• **Chaff**: Dropping metal strips from an aircraft to create confusing radar reflections.

• **Decoys**: Employing small flying objects that look like full-size aircraft on a radar screen.

• **Anti-radiation missiles**: Destroying enemy radar by homing in on radio signals they transmit.

• **Stealth**: Manufacturing military aircraft that produce little or no radar reflection.

⭢ Joining one spacecraft with another in space is called docking, and radar is needed for this maneuver to measure distances and speed of approach. This computer-generated image shows the Space Shuttle *Atlantis* docked to the Russian space station *Mir*, which orbited Earth from 1986 to 2001.

capsules, and unmanned Progress supply craft have docked with the Russian *Mir* space station and with the International Space Station.

Docking is a difficult maneuver. In space, without any nearby landmarks by which to judge distance and speed, it is almost impossible to determine how far away a spacecraft is or how fast it is moving. A crew onboard the *Mir* space station in 1997 discovered this problem when they were docking an unmanned supply craft with their space station without the use of radar. The craft, controlled by a cosmonaut in *Mir*, approached the station too fast and crashed into it.

Radar is usually used for all docking maneuvers. It provides accurate measurements of the distance between spacecraft and their closing speed. Linked to a spacecraft's guidance system, it can carry out docking automatically. When the Apollo lunar excursion modules descended from lunar orbit to the Moon's surface during the Apollo landings of the 1960s and 1970s, the descent was controlled by radar. Landing radar kept the guidance computer constantly updated with data on the spacecraft's altitude and rate of descent.

Radar's Many Uses

Many home security systems have motion detectors that sense when someone is moving around in a room. Some of these detectors work by picking up the heat of the person's body, but others use radar. They flood the room with microwaves that bounce back to the detector. If someone enters the room, the steady pattern of reflections is disturbed, and an alarm sounds.

Some of the cameras used to monitor the speed of vehicles on highways work by radar. The radar system measures the speed of the vehicles. If a vehicle is moving faster than the speed limit, the radar system triggers a camera that photographs the car, including its license plate, so it can be traced to the speeding owner.

Archaeologists use a variety of methods to map structures underground and ground-penetrating radar is one of these methods. Radar can probe down to 33 feet (10 meters) deep and show buried features of ancient buildings such as walls and floors.

SEE ALSO:
- Air Traffic Control
- Communication
- Space Race • World War II

HOW SPACE PROBES MAKE MAPS

Space probes use radar to make maps of planets. As they orbit the planet, they fire radar pulses at the planet's surface. This measures the distance between the spacecraft and the planet and thereby builds up a map of its surface shapes. The bigger a radar antenna is, the more detailed a map it can make, but spacecraft can only be fitted with small antennae. Clever signal processing enables these small antennae to work as if they are much bigger. A number of radar reflections are put together in a series as if they had come from one big antenna instead of a small antenna moving along. This is called synthetic aperture radar (SAR). The Magellan space probe mapped Venus using SAR. Remote sensing satellites in Earth orbit also use SAR to produce detailed images of our planet.

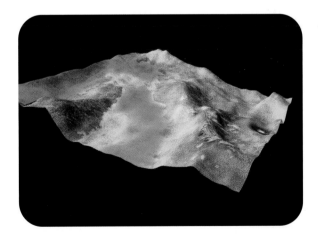

A SAR image shows lowlands, ridges, hills, and (right) an impact crater on the surface of Venus.

Relativity, Theory of

In the early 1900s Albert Einstein (1879–1955) produced two theories that caused a revolution in science. Together, they are known as the theory of relativity.

The first of Einstein's relativity theories is his Special Theory of Relativity of 1905. This states that:

(1) The laws of physics are the same for people moving at different speeds.

(2) The speed of light is always the same for all observers, no matter how fast they are moving.

These two simple principles produce some surprising effects when speeds close to the speed of light are involved.

There is no such thing as absolute rest anywhere in the universe. Someone standing still on Earth is actually moving because Earth is spinning. The planet also is orbiting the Sun. The Sun orbits the center of the Milky Way (our galaxy), and the Milky Way moves through space among billions of other galaxies that also are moving. Any observer can say that he or she is at rest and everything else in the universe is moving in relation to—or relative to—him or her. This phenomenon gives relativity its name.

One of the most surprising effects of relativity is that times and lengths depend on who measures them. Imagine that one person stays on Earth while her twin goes on a long spaceflight almost as fast as the speed of light. The person on Earth would see the spacecraft become shorter as it accelerates toward the speed of light. This is called Lorentz contraction. Also, time runs more slowly as speed increases. When the astronaut twin returns to Earth, she would be younger than her earthbound twin. This is known as the "twins paradox."

The Special Theory of Relativity also predicts that as something goes faster, it becomes more and more difficult to make it go even faster. The more energy it has, the more inertia it has. Einstein also showed that energy

⟲ The twins paradox illustrates Einstein's theory that if two bodies at the same point accelerate to different speeds and then meet again, they will find that a different amount of time has elapsed for each of them.

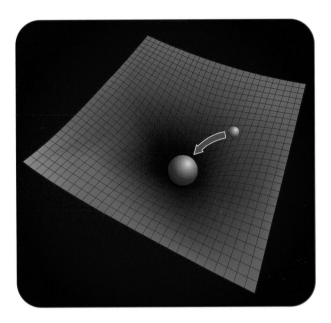

🎧 If space were a stretched rubber sheet, a heavy ball on the surface would curve the sheet by sinking into it. According to Einstein's General Theory of Relativity, gravity occurs because smaller moving objects curve toward the larger mass.

and mass are linked. He produced his famous equation, $E=mc^2$, to show how they are related. In this equation, "c" is the speed of light, a huge number, so even a tiny mass (m) is equivalent to an enormous amount of energy (E).

In 1915, Einstein produced a new theory, his General Theory of Relativity, which added the effect of gravity to the Special Theory. Nobody had been able to explain how gravity works. Einstein imagined that space is flat and all objects dent it, like heavy weights sitting on a stretchy sheet. In other words, mass bends space. The paths of moving objects curve toward massive objects, simply because space itself is curved there. Light also is bent by curved space.

TIME CORRECTION

Global Positioning System (GPS) navigation satellites fly around the world at 7,000 miles per hour (11,300 kilometers per hour). According to special relativity, their onboard clocks should run more slowly than clocks run on Earth. According to general relativity, the curving of space caused by Earth has the opposite effect. It makes the clocks run faster than clocks on Earth. The result of these two counteracting effects is that the satellite clocks run 38-millionths of a second faster every day than clocks on Earth. It is a tiny error, but GPS clocks have to be more than 1,000 times more accurate than this. So satellite clocks are deliberately set to run slow before they are launched, so that they are correct in orbit.

Relativity is not just a strange set of ideas and mathematics. Its effects are real. Observations and experiments have confirmed many of its predictions. Spacecraft designers and planners of space missions have to allow for them.

SEE ALSO:
- Einstein, Albert • Global Positioning System • Spaceflight

Ride, Sally

Date of birth: May 26, 1951.
Place of birth: Los Angeles, California.
Major contribution: First American woman to reach space.
Awards: Induction into National Women's Hall of Fame and the Astronaut Hall of Fame; Jefferson Award for Public Service; Von Braun Award; Lindbergh Eagle; NCAA's Theodore Roosevelt Award; NASA Space Flight Medal (twice).

After graduating from high school in Los Angeles, California, Sally Ride went on to attend Stanford University in California. She graduated in 1973 with degrees in physics and English. Deciding to focus on astrophysics, Ride earned her master's degree in 1975 and her doctorate in 1978, both from Stanford.

In January 1978, Ride was one of six women chosen by NASA for astronaut training. Her first chance to fly in space came in 1983. On June 18, she joined four other astronauts on STS-7 aboard the Space Shuttle *Challenger*. Ride's main task was to work the Space Shuttle's robot arm. This mission was the first to use the robot arm to deploy and to retrieve a satellite. STS-7 flew for six days before returning to Earth.

Ride's second mission, STS 41-G, took off on October 5, 1984. Once again she flew in the Space Shuttle *Challenger*. This mission lasted eight days, and Ride worked the robot arm to deploy a satellite. The seven-member crew also carried

🎧 Sally Ride, shown monitoring control panels on the Space Shuttle flight deck, was a Space Shuttle mission specialist in 1983.

out experiments. This flight made Ride the first American woman to fly twice in space. Fellow astronaut Kathryn Sullivan became the first American woman to walk in space.

Ride was assigned a further Space Shuttle flight in 1985 and began preparing for a launch the next year. That mission was canceled when, in January 1986, *Challenger* exploded shortly after takeoff. The *Challenger* disaster caused NASA to ban further Space Shuttle flights until the cause of the explosion could be determined. Ride was chosen to sit on the commission that investigated the accident.

After that work was complete, she transferred to NASA headquarters in Washington, D.C., where she worked on long-range planning for the agency. Her *Ride Report*, issued in 1987, recommended using the technology of space exploration to study conditions on Earth. This *Mission to Planet Earth*, as it was called, has been undertaken by NASA. Much of the research focused on the issue of climate change. Another Ride recommendation was to begin planning for a mission to Mars. At the time, NASA did not pursue this plan, but instead focused its work on the International Space Station.

In 1987, Ride left NASA to accept a position at the Stanford University Center for International Security and Arms Control. Two years later, she joined the faculty of the University of California at San Diego, where she taught and carried out research in physics. For many years, Ride also directed the California Space Science Institute, although she left that post to focus on research, teaching, and her many other activities.

Over the next twenty years, Ride served on several government committees involved with space and technology. When the Shuttle *Columbia* exploded in 2003, NASA launched a new investigation. As with the *Challenger* incident, Ride was a member of the commission studying that accident.

Ride also dedicated herself to promoting interest in science and space exploration among young people, especially girls. She wrote several children's books on space and took an active role in other efforts to build the popularity of space exploration. In 2001, she founded her own company, Sally Ride Science, to motivate girls and young women to pursue careers in science, math, and technology.

On June 21, 2003, at the Kennedy Space Center in Florida, Sally Ride was inducted into the U.S. Astronaut Hall of Fame. Alongside her on the platform are former astronauts, all members of the Hall of Fame.

SEE ALSO:
• Astronaut • *Challenger* and *Columbia* • NASA • Space Shuttle

Rocket

A rocket is a type of jet engine. It has three main parts: its structure, a guidance system, and a propulsion system. The structure is the basic frame of a rocket. The guidance system keeps the rocket on course. The propulsion system comprises the rocket's engines. A rocket carries a payload—the spacecraft that is launched by the rocket. Rockets contain everything they need to work and do not need to take in air.

The History of Rockets

It is not known who built the very first rocket, but it is believed that some form of rocket was made in China sometime before 1200, perhaps as long ago as 200 B.C.E. Early rockets were like fireworks, propelled by a mixture of chemicals called black powder. The first recorded use of rockets in war was at the Battle of Kai-Feng in 1232, when a Mongol attack on the Chinese city of Kai-Feng was fought off with the help of rockets.

In the eighteenth century, British forces came under rocket attack in India. They were so impressed with the rockets that a British artillery officer, Sir William Congreve, developed solid fuel military rockets for the British army. Congreve's rockets were used during the Napoleonic Wars in Europe.

KONSTANTIN TSIOLKOVSKY (1857–1935)

Konstantin Tsiolkovsky, spaceflight's earliest pioneer, was a Russian schoolteacher who wrote scientific studies of rockets and space travel. In 1898, Tsiolkovsky wrote an article entitled "Investigating Space with Rocket Devices" that presented the scientific principles of spaceflight. Over the next thirty or so years, Tsiolkovsky wrote a series of scientific and mathematical studies of rocket engines, rocket fuels, spacecraft in orbit, space stations, and even spacesuits.

The many honors he was awarded included a lifetime pension from the Soviet state that enabled him to retire from teaching and work full-time on spaceflight. All of Tsiolkovsky's work was theoretical. He never carried out any experiments with rockets, but rocket scientists and engineers in the Soviet Union and other countries have studied Tsiolkovsky's work.

🔊 Konstantin Tsiolkovsky with a model of an early Russian rocket.

These early rockets were very inaccurate. Like firework rockets, they relied on a long stick to keep them pointing in the right direction and to stop them from tumbling. This was improved upon in 1844, when William Hale designed a rocket that spun as it flew through the air. The spinning kept it flying straight and true. These early military rockets were overtaken by advances in artillery.

In 1898, a Russian schoolteacher named Konstantin Tsiolkovsky wrote the first serious study of spaceflight using rockets. In the 1920s, rocket research flourished in several countries. In the United States, Robert H. Goddard launched the first liquid-fuel rocket. Amateur rocket scientists in Germany began developing rockets that led to the first modern rocket-propelled weapons in World War II.

The Engine and Fuel

The force produced by a rocket engine is called thrust. The upward force of thrust must be greater than the downward force of gravity if a rocket is to take off. One measure of an engine's power and efficiency is its thrust-to-weight ratio. Rockets have the highest thrust-to-weight ratios of all engines.

Most rockets work by means of chemical reactions, combining two chemicals, or propellants: a fuel and an oxidizer. The oxidizer, consisting of oxygen or a

⮑ A rocket must produce huge upward thrust to get into space. It does this by producing a high-pressure jet of hot gases.

chemical containing oxygen, is needed to burn the fuel. Burning produces hot gases that expand rapidly and rush out of the engine nozzle at high speed. According to Newton's third law of motion, the gas jet pushes against the rocket, and the rocket pushes back against the gas jet with the same force. The result is that the rocket is thrust in one direction as the gas flies in the opposite direction.

Other types of rockets produce thrust from a high-pressure jet of water, air, steam, or another gas. Ion engines produce thrust by accelerating electrically charged gas particles.

One way to make a rocket accelerate faster, go farther, or carry a heavier payload is to make it lighter. For this reason, large rockets usually are built as a series of rockets standing on top of each other. The individual rockets are called stages. When each stage uses up its fuel, it falls away and the next stage fires. This enables the rocket to get rid of unnecessary weight that would slow it down.

Solid-Fuel Rockets

Solid-fuel rockets are the simplest and oldest type of rockets. Aircraft have been armed with solid-fuel rockets since World War I, when they were used to attack airships and observation balloons. Rockets fired from one aircraft at another aircraft are called air-to-air rockets. The airships and balloons were filled with hydrogen, which burned if a flaming rocket flew into it. The problem was that the planes of the day also were made of flammable materials, so firing rockets from them was dangerous. The rockets also were inaccurate and rarely hit their targets.

Small air-to-air rockets were used again in World War II. They enabled fighters to attack bombers without coming within range of the bombers' guns. These small rockets were unguided—they were aimed simply by pointing the

⊂ The Space Shuttle has two solid rocket boosters (SRB) that are strapped to its external fuel tank. The SRB are discarded about 2 minutes after liftoff, and they fall back to Earth to be retrieved and reused.

ROBERT H. GODDARD (1882–1945)

Robert Hutchings Goddard was an American scientist and inventor who developed the modern liquid-fuel rocket. He received patents for a liquid-fuel rocket and a two-stage rocket in 1914. In 1919, Goddard wrote a paper called "A Method of Reaching Extreme Altitudes," in which he talked about sending a rocket to the Moon. He was ridiculed at the time for even suggesting such a crazy idea. In 1926, Goddard succeeded in building and launching the first liquid-fuel rocket. Powered by gasoline and liquid oxygen, the small rocket rose to a height of 41 feet (12 meters). Goddard went on to build bigger and more powerful rockets. Some of them climbed higher than 9,000 feet (2,740 meters) and went faster than the speed of sound. Goddard was the first person to steer a rocket by using vanes in the rocket exhaust, and he designed the first gyroscopic systems for guiding rockets. NASA's Goddard Space Flight Center is named in his honor.

↻ Robert Goddard displays his liquid oxygen-gasoline rocket before its successful launch in 1926.

whole plane. In the 1950s, air-to-air rockets were replaced by guided missiles.

Solid-fuel rockets are used to help launch spacecraft. Space launch rockets are liquid-fuel rockets, but they can be made more powerful by strapping solid-fuel rockets around them. The solid rockets provide extra power for liftoff. Extra rockets used like this are called boosters. The Space Shuttle is launched with the help of two solid rocket boosters (SRB). They burn powdered aluminum fuel with ammonium perchlorate oxidizer. The propellants are mixed as liquids and then set hard in a mold. A hole runs through the center of the rocket. When the propellants are ignited, they burn from the inside out. Once solid-fuel rockets have been lit, they cannot be turned off.

↺ The manned spacecraft *SpaceShipOne* uses a hybrid rocket motor. It has made several sub-orbital space-flights.

Liquid Fuel and Other Propellants

Controlled spaceflight needs a rocket in which the power can be varied and turned on and off. Liquid-fuel rockets can be controlled in this way. Liquid-fuel rockets are more complicated than solid rockets, because piping, valves, and pumping systems are needed to move the liquid propellants from their storage tanks to the engines. A type of kerosene called RP-1 (Refined Petroleum-1) is a commonly used liquid rocket fuel.

Unlike RP-1, some liquid propellants have to be kept very cold. Hydrogen and oxygen are common rocket propellants. They are normally gases, but they can be packed into very small tanks by changing them into liquids. Hydrogen becomes liquid below a temperature of -423°F (-253°C). Oxygen becomes liquid below -298°F (-183°C). Liquid oxygen also is called LOX. Propellants that have to be kept super-cold are known as cryogenic propellants. They are not suitable for most military rockets and missiles because it is difficult to keep them sufficiently cold for long periods, and military equipment always must be kept ready to use. Instead, cryogenic propellants are used for civilian spaceflight, such as the Space Shuttle missions, because they are highly efficient, yielding a lot of power per gallon.

Some small rockets use propellants that ignite as soon as they meet. These are called hypergolic propellants. Rocket engines that use hypergolic propellants can be very simple and reliable, because they do not need complicated ignition systems. Small rockets called thrusters use hypergolic propellants.

Strange materials have been used as rocket propellants. The *Mythbusters* television program, which aims to prove or disprove myths, built a working rocket fueled by a salami. *SpaceShipOne*, the first privately funded manned space plane and winner of the Ansari X-Prize, burns rubber as its fuel. The rubber is solid, and the oxidizer, nitrous oxide, is liquid. A rocket like this, with a mixture of solid and liquid (or gas) propellants, is called a hybrid rocket.

WERNHER VON BRAUN (1912–1977)

Wernher von Braun was the German-born rocket scientist and engineer who created the giant Saturn V rockets that landed U.S. astronauts on the Moon. After studying engineering, he earned a doctorate in physics at the University of Berlin in Germany. He joined the Society for Space Travel, which was led by the rocket scientist Hermann Oberth. Von Braun's work in the society was noticed by leaders of the German army, who hired him to develop missiles during World War II. Von Braun's team at Peenemünde in northeast Germany developed a series of rockets, including the famous V-2. The V-2 could hit targets up to about 185 miles (300 kilometers) away.

At the end of the war, the United States and Soviet Union captured unused V-2s as well as some of the scientists and engineers who had worked on them. In 1945 von Braun surrendered to U.S. troops and went to work in the United States. The first rockets built in the United States (and the Soviet Union) in the 1950s were based on von Braun's V-2. Braun led a team that developed a series of rockets and missiles, including the Redstone, Jupiter-C, Juno, and Pershing. When NASA was formed, von Braun went to work there and developed the Saturn I, IB, and V rockets. He founded the National Space Institute to promote public understanding of spaceflight. Von Braun also wrote several popular books on spaceflight and gave talks on the subject. He received numerous awards in recognition of his work.

☾ Wernher von Braun was director of NASA's Marshall Space Flight Center from 1960 to 1970.

Steering and Braking

There are several ways of steering a rocket or rocket-powered spacecraft. One way is to use swiveling fins, like an airplane's control surfaces, in the atmosphere. A rocket must be traveling fast before its fins begin to work, because they only work when air is flowing over them very quickly. Other rockets have swiveling vanes in the rocket exhaust. When the vanes swivel, they deflect some of the engine's exhaust jet. The entire jet can be deflected by swiveling the engine itself or just the nozzle. Most modern rockets have swiveling engines, also called gimbaled engines. The Space Shuttle's main engines are gimbaled.

A rocket or spacecraft also can be turned or steered by means of thrusters.

When the Space Shuttle's solid rocket boosters fall away, thrusters push them away from the spacecraft. The Space Shuttle uses forty-four thrusters in its nose and tail for attitude control when flying in space.

Rockets are used for braking as well as steering. Braking rockets also are called retro-rockets. When an orbiting spacecraft is ready to land, it fires off rockets in the direction in which it is traveling. The thrust slows the spacecraft, and gravity begins to pull it down.

⟲ The North American Aviation X-15 rocket plane made nearly 200 flights between 1959 and 1968, reaching a top speed of 6.7 times the speed of sound and a maximum altitude of 354,200 feet (107,960 meters).

➲ The C-130 Hercules aircraft that travels with the U.S. Navy's Blue Angels display team sometimes uses JATO to get airborne. The JATO rockets are visible on the side of the plane.

The Soyuz spacecraft uses retro-rockets for landing. It fires retro-rockets just before it touches down on the ground to cushion its landing.

Rocket-Powered Airplanes

Rocket-powered airplanes are rare, because the propellants that power them are often poisonous, explosive, or have to be kept super-cold. The German company Messerschmitt built a rocket-powered fighter, the Me163 Komet, in the 1940s. It could climb amazingly fast, but was could only stay airborne for about 8 minutes.

Rocket-powered planes have been used for research in high-speed flight. On October 14, 1947, the first supersonic flight was made in the rocket-powered Bell X-1 aircraft with Chuck Yeager at the controls.

Rockets are sometimes used to help heavy aircraft take off. This is called rocket assisted takeoff (RATO) or jet assisted takeoff (JATO). The solid-fuel rockets used for this are called JATO bottles because they look like big bottles.

Other Uses

Small rockets are used for a variety of purposes. Fighter pilots sit in rocket-powered ejection seats. If a pilot has to leave an aircraft in an emergency, rockets blast the seat clear of the aircraft. Rocket flares for signaling an emergency at sea use a rocket to launch a bright flare, which may then descend slowly by parachute. Scientists use small rockets called sounding rockets to carry instruments into the upper atmosphere. Lightning researchers also use rockets to trigger lightning for study.

> **SEE ALSO:**
> • Apollo Program • Bell X-1
> • Ejection Seat • Engine • Fuel
> • Jet and Jet Power • Spaceflight
> • Space Shuttle

Satellite

A satellite is an object in space that orbits another body, such as a planet. The Moon is a natural satellite of Earth. An artificial satellite is a small spacecraft sent into orbit from Earth.

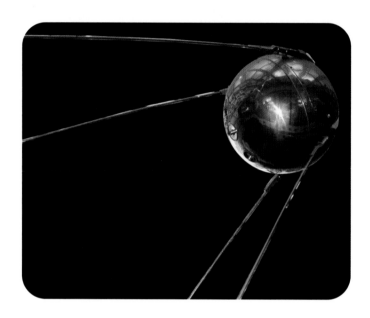

🔊 *Sputnik 1* was the first satellite that was successfully sent into orbit around Earth. It was launched in 1957 and stayed in orbit for about three months.

Artificial satellites are used for communication, weather forecasting, navigation, military surveillance, and scientific research. A satellite remains in orbit because of the gravitational pull of the larger body around which it travels. Any spacecraft orbiting Earth is technically a satellite, although manned spacecraft are not usually referred to as satellites. The International Space Station is the biggest artificial satellite. The Hubble Space Telescope also is a satellite. There are about 2,500 satellites orbiting Earth, and others have been placed in orbit around the Moon, Venus, and Mars.

The First Satellites

The idea of "artificial moons" orbiting Earth was put forward by a few scientists and science-fiction writers in the early 1900s. In 1945, British science fiction writer Arthur C. Clarke suggested in a magazine article that three satellites orbiting Earth could act as relay transmitters for worldwide communications. The visionary Clarke now has a satellite orbit named after him.

By the 1950s, there were rockets capable of launching satellites, although these were being developed primarily as military missiles. The American Rocket Society and the National Science Foundation both suggested using satellites for the scientific study of space. These groups proposed that, to mark International Geophysical Year (1957–1958), the United States should launch a science satellite.

The Soviet Union announced that it also would launch a satellite, but few people in the West took this claim seriously. So it was a great surprise when, on October 4, 1957, the Russians launched the world's first satellite, *Sputnik 1*. Orbiting Earth every 96 minutes, *Sputnik 1* caused a sensation. An even greater surprise followed on November 3, 1957, with the launch of *Sputnik 2*, which was bigger still and carried a dog named Laika. The United

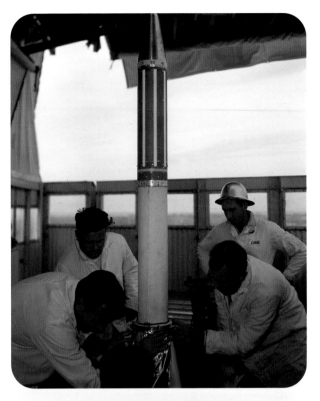

⤳ The first satellite launched by the United States was *Explorer 1*, seen here being installed on its launch vehicle in January 1958.

States launched its first satellite, *Explorer 1*, early in 1958. The "space race" had begun in earnest.

The Soviet Union launched hundreds of Cosmos satellites and also Molniya communications satellites, but seldom released much information about them. U.S. space launches were more public, and satellites were usually designated by their function: scientific, weather, communications, navigation, or Earth observation. Only satellites for military use were kept secret.

Since the 1960s, France, China, Japan, Britain, India, and Israel have launched satellites with their own launch vehicles. Other nations have hired launches to put satellites into space. Once front-page news, satellites are now routine, with many commercial and multinational launches each year.

Satellite Basics

Satellites vary in size from a few pounds to many tons. Some remain in orbit for only a few weeks, while others have an expected lifetime of hundreds of years. They are packed with scientific instruments, usually miniaturized to save weight and space. Manufacturing is closely monitored and takes place in germ-free conditions. All systems are thoroughly tested—once launched, satellites must continue to work under remote control for long periods of time.

⬆ A photo taken from the Space Shuttle *Challenger* shows the spacecraft's cargo bay open as it releases a satellite into orbit in 1984. This satellite was retrieved by *Columbia* in 1990.

SATELLITE LAUNCHERS

A satellite can be carried into space in the cargo bay of a Space Shuttle, or it may be blasted into space on top of a multistage rocket or launch vehicle. An expendable launch vehicle, or ELV, is used only once. An ELV has two or more booster stages; each stage falls away when its engine burn is completed. The final stage sends the payload (the satellite) into orbit.

The launch vehicle Pegasus is itself launched from beneath a converted Lockheed L-1011 aircraft. The launcher can place a satellite weighing up to 970 pounds (440 kilograms) into near-Earth orbit. Pegasus launched the *Solar Radiation and Climate Explorer* and the *Galaxy Evolution Explorer* satellites in 2003.

A larger satellite, or a satellite intended for high orbit, requires a more powerful launcher, such as a ground-launched Athena or Delta rocket. Delta rockets have launched many satellites since the 1960s, including *TIROS, Nimbus, ITOS,* and *Landsat* satellites. The big Delta IV can launch a payload of 50,800 pounds (23,070 kilograms) into low-Earth orbit.

⋒ A huge Delta IV rocket stands at Cape Canaveral, ready to carry an observation satellite into space in 2005.

Most satellites are fitted with panels of solar cells that convert energy from sunlight into electricity. They use solar energy to power instruments and the communication systems that send data, such as video images, back to Earth.

A satellite is directed by remote control from mission centers on the ground, where scientists monitor its orbit, send instructions, and receive data from the satellite's instruments. Under command from Earth, a satellite can be shifted in orbit by firing small thruster rockets.

A satellite may be in constant communication with mission control. If it is in a low orbit, it may be contacted only

A polar-orbiting satellite is prepared for launch in 2000. The satellite joined the polar-orbiting operational environmental satellite (POES) program, which provides data about the global climate and weather.

when it passes overhead. Each pass may last just 10 minutes, but the satellite may fly by ten or twelve times a day, depending on its orbit. Some satellites are visible at night as they pass overhead.

A satellite that has malfunctioned, or has come to the end of its operational life, is normally shut down and then allowed to burn up as it reenters the atmosphere. Some defective satellites, however, have been picked up by Space Shuttle astronauts for repair.

Satellite Orbits

A satellite's orbit depends on the task for which it is designed. Most satellites are launched in the same direction as Earth is spinning, and this is called a prograde orbit. To launch in the opposite direction, like throwing a ball into the wind, requires more booster power and fuel.

Scientists choose various orbits for their satellites, depending on the location of the launch site and the task of the satellite. Orbits fall into three types: high geostationary orbit, Sun-synchronous polar orbit, and low orbit.

A high geostationary orbit keeps a satellite always in the same position with respect to Earth. The satellite makes one orbit in the same period of time as Earth makes one rotation (23 hours, 56 minutes, 4.09 seconds). To do this, it must orbit at a height of about 22,300 miles (about 35,900 kilometers) above Earth's surface. By orbiting in tandem with Earth, the satellite appears stationary, or synchronous (in time), with respect to the rotation of the planet.

A Sun-synchronous polar-orbiting satellite travels above the North and South poles. It flies at a height of about 540 miles (about 870 kilometers) and passes the Equator and each of Earth's latitudes at the same time each day. Being Sun-synchronous means the satellite passes overhead at the same solar time through the year, so it can transmit data (on weather, for example) at consistent times. Its data can be compared year by year.

Low-orbiting satellites fly at a height of 200–300 miles (320–485 kilometers). A low orbit requires the least rocket

NOTABLE SATELLITES

Name	Country	Date	Achievement
Sputnik 1	Soviet Union	1957	First artificial satellite.
Sputnik 2	Soviet Union	1957	First satellite to carry a passenger (a dog).
Explorer 1	United States	1958	First U.S. satellite.
Explorer 6	United States	1959	First photos from space.
Tiros 1	United States	1960	First weather satellite.
Transit 1B	United States	1960	First navigation satellite.
Telstar	United States	1962	First communications satellite.
Landsat	United States	1972	Series of Earth-survey satellites.
Pegsat	United States	1990	First airplane-launched satellite.
Mars Global Surveyor	United States	1997	Orbited and mapped Mars.
Envisat	Europe	2002	Series of satellites measuring global warming.
Aura	United States	2004	Satellites studying the ozone layer.
Themis	United States	2007	Five satellites launched simultaneously to study Earth's magnetosphere.

power and is often chosen for observatory satellites, such as the Hubble Telescope. Hubble orbits Earth at a height of about 375 miles (600 kilometers), making one orbit every 97 minutes.

Some orbits are circular, while others are elliptical (egg-shaped). The length of time a satellite takes to make one orbit is called its orbital period.

A satellite's initial velocity is high enough to counter the force of gravity and keep it in orbit, but friction (from Earth's atmosphere and from the Sun's energy) gradually slows the satellite's speed. Its orbit begins to decay. Eventually, as the satellite descends into the thicker layers of the atmosphere, it burns up or breaks up.

Military Satellites

Military satellites include spy or reconnaissance satellites. These satellites are fitted with scanning devices and cameras that can detect objects on the ground. Some of these objects may be as small as a truck hundreds of miles below the spacecraft. Spy satellites also can detect missiles being fired. The first military satellite able to detect missile launches was *Midas 2*, launched by the United States in 1960. Early spy satellites took photographs on film that were returned to Earth in small capsules that landed by parachute. Modern spy satellites are equipped with digital imaging systems, and they relay their images directly from space.

A number of countries have military satellites. Military navigation satellites are used by aircraft, submarines, surface ships, and land vehicles. Anti-satellite weapons, known as "killer" or "suicide" satellites, are designed to track, locate, and destroy other satellites or orbital weapons systems.

Navigational Satellites

Navigational satellites are very useful pieces of space equipment. They provide the Global Positioning System (GPS) network, which enables pilots, sailors, drivers, and hikers to fix positions almost anywhere on the globe with pinpoint accuracy. Developed by the U.S. Department of Defense as NAVSTAR (Navigation Satellite Timing and Ranging Global Positioning System), the GPS uses at least twenty-four satellites to make sure that at least four are always within the line of sight of a navigator on the ground or ocean. One early navigation satellite, *Transit 4A*, was the first satellite to carry a small nuclear power plant.

Earth Observation and Weather Satellites

Earth observation (environmental) satellites are used to monitor changes in the environment such as melting ice caps, deforestation, and desertification. Earth observation satellites are normally launched into Sun-synchronous polar orbits so that they can survey the entire globe. They can scan for minerals, water, and other resources and record land use

ASAT MISSILE SYSTEMS

The United States and the Soviet Union tested anti-satellite (ASAT) missile systems back in the 1980s. In 1985, a U.S. F-15 fighter fired a missile that flew into space and destroyed a U.S. solar observatory satellite orbiting 375 miles (600 kilometers) from Earth. After this one success, the ASAT project was abandoned, partly because of concerns that such missiles violated the 1967 Outer Space Treaty. The treaty requires nations to refrain from placing weapons into space—such as nuclear warheads, lasers, and other high-energy weapons—that could be used to destroy satellites or aimed at ground targets. In 2007, China claimed to have test-fired a missile that destroyed an obsolete weather satellite, raising a new debate about ASAT usage. As increasing numbers of satellites are launched, the question of how to regulate ASAT systems remains unresolved.

⟳ This ASAT missile was successfully released to destroy a satellite during a 1985 test.

in wilderness and cities. Space cameras provide images from which mapmakers create accurate maps. They even can give computer users instant images of their own location over the Internet.

Weather satellites have revolutionized meteorology. They provide the daily TV weather images, and they alert forecasters to developing global weather situations, such as hurricanes. The National Oceanic and Atmospheric Administration (NOAA) runs a national weather service from satellite data provided by the National Environmental Satellite Data and Information Service.

Short-range weather forecasting uses data from geostationary operational environmental satellites (GOES). Long-range weather forecasts use data from polar-orbiting operational environmental satellites (POES). NOAA also operates a search-and-rescue satellite-aided tracking system, known as SARSAT,

which can locate a person in trouble at almost any location on the planet.

Communications Satellites

Telecommunications providers use communications satellites (comsats), which function as relays for telephone, radio, and television signals. The first satellite able to relay a voice signal was launched in 1960; *Telstar* was the first real communications satellite, launched in 1962. *Syncom 3*, launched in geostationary orbit in 1963, relayed the 1964 Tokyo Olympics to U.S. viewers, the first television pictures sent across the Pacific Ocean. *Intelsat 1*, also known as "Early Bird," relayed TV signals across the Atlantic in 1965. Satellites launched for commercial companies revolutionized satellite and cable TV. They made satellite television possible—today there are hundreds of channels, and live coverage of events is transmitted all over the world. Groups of satellites also provide worldwide phone networks.

Military comsats such as the U.S. Milstar system (launched in 1994) provide secure communications that cannot be blocked. In the 1960s, the Russians launched a series of Molniya comsats into elliptical, 12-hour orbits, with perigees (low points) of no more than a few hundred miles and apogees (high

⮃ Geostationary operational environmental satellites (GOES) provide views of Earth that help forecasters accurately predict emergency weather conditions. This GOES image shows Hurricane Andrew over the Gulf of Mexico in 1992.

🎧 *Chandra*, named for a leading Indian astrophysicist, Subrahmanyan Chandrasekhar, is one of the largest satellites ever. It carries eight mirrors to focus X-rays from distant objects, a high-resolution camera, and a spectrometer to measure the amount of energy in the X-rays.

points) of up to 25,000 miles (40,230 kilometers). This kind of orbit is now called a Molniya orbit. Less rocket power is needed to put a satellite into this orbit than into a high geostationary orbit.

Scientific Satellites

Science satellites carry out a range of tasks to observe objects and phenomena in space. They have transformed many scientists' views of the universe. Science satellites' instruments often measure radiation in various forms. *IRAS* (*Infrared Astronomical Satellite*) was launched in 1983. In its ten-month lifespan, it discovered 20,000 galaxies (including a new kind called a starburst galaxy), 130,000 stars, and a comet. In 1999, the Space Shuttle *Columbia* launched an X-ray observatory named *Chandra*. It has an unusual elliptical orbit that brings it to

SPACE JUNK

Old satellites and leftover pieces of satellite launch vehicles end up as trash drifting in space. Space junk is heaviest at a height of around 530 miles (850 kilometers), where most satellites orbit. After a half-century of space launches, there is now a lot of space junk. Scientists have recorded at least 11,000 objects larger than 4 inches (10 centimeters) in diameter. The junk includes used-up rocket stages, tools lost by astronauts, and lumps of solidified fuel. Space junk is a potential hazard, since a lump of fist-size debris, traveling at more than 21,000 miles per hour (33,800 kilometers per hour), can make a serious hole in an expensive spacecraft.

within 6,000 miles (9,650 kilometers) of Earth and then swings out to 86,000 miles (about 138,400 kilometers)—about a third of the way to the Moon. Each orbit takes 64 hours. Being so far out means *Chandra* keeps clear of the belts of charged particles that surround Earth and so provides astronomers with longer periods of clear observation time.

SEE ALSO:
- Gravity • Rocket • Space Probe
- Space Race

Shepard, Alan

Date of birth: November 18, 1923.
Place of birth: East Derry, New Hampshire.
Died: July 21, 1998.
Major contribution: First American in space.
Awards: Congressional Space Medal of Honor; two NASA Distinguished Service Medals; NASA Exceptional Service Medal; Navy Distinguished Service Medal; Navy Distinguished Flying Cross; several other trophies, medals, and honorary degrees.

After Alan Shepard's first trip in an airplane, in his early teens, he became fascinated by flying. He often visited the local airport, doing odd jobs in the hope of a plane ride. Shepard attended the U.S. Naval Academy at Annapolis, Maryland. Graduating in 1944, he served on a destroyer during the last year of World War II. After the war, Shepard became a navy pilot, and, in 1950, he became a test pilot.

In 1959, the National Aeronautics and Space Administration (NASA) began recruiting the first American astronauts. The agency sent invitations to the top 110 test pilots. Shepard entered the program and soon after was named as one of the seven Project Mercury astronauts. After two years of training, Shepard was chosen to be the first American in space.

Weather and technical problems delayed Shepard's flight, but the launch finally took place on May 5, 1961. Takeoff was smooth, and the flight was brief. Shepard never reached orbit—he simply went up and, about 15 minutes later, splashed down. Splashdown and recovery were successful. The launch and the recovery were covered live by television, and Americans greeted Shepard as a hero.

After the celebrations were over, Shepard returned to NASA. In 1964, before he could make another space-flight, he developed a serious problem in his inner ear. Fluid buildup would, from time to time, cause him to lose his balance and feel nauseous. NASA grounded Shepard, and he took on the alternative job of chief of astronaut operations. After several years, he decided to have surgery to try to correct his ear problem. The 1969

↶ Alan Shepard is seen here being recovered by a helicopter after splashdown in May 1961.

➲ Alan Shepard was photographed on the Moon with a transporter used for carrying equipment and samples. Shepard and fellow astronaut Edgar Mitchell spent more time on the Moon (33 hours) than any other Apollo astronauts.

operation was a success, and soon after, Shepard was cleared for spaceflight again.

Shepard was ready to achieve his dream of flying to the Moon. He was named commander of Apollo 14, teamed with Edgar Mitchell and Stuart Roosa. On January 31, 1971, Shepard returned to space. Five days later, he and Mitchell landed on the Moon. They spent more than a day on the Moon's surface, where they collected a large sample of moon rocks and carried out several experiments. Having completed their mission, the astronauts returned safely to Earth.

Shepard continued as chief of astronaut operations until he retired from NASA three years later. He began working in business, where he was successful in several ventures. In 1984, Shepard, five other Mercury astronauts, and the widow of a seventh astronaut formed a foundation that gave scholarship money to students interested in science and engineering. Shepard led the foundation until stepping down in 1997. He died the following year.

SEE ALSO:
- Apollo Program • Astronaut
- NASA

GOLF ON THE MOON

On Apollo 14, Shepard decided to indulge his passion for golf. Before the flight, he had a NASA worker cut the head off a golf club and attach a device that could be used to connect it to a Moon exploration tool. Before launch, Shepard stuffed the club head and two golf balls into a sock and hid them in his spacesuit. At the end of his Moon walk, he surprised NASA officials by attaching the club head and smacking the two balls. All this took place on live television. Although the first ball did not go far, the second—Shepard announced—traveled for "miles and miles."

Shock Wave

A shock wave in air is a sudden, huge rise in air pressure. Shock waves affect the flight of high-speed aircraft and spacecraft through the atmosphere.

Everyone who has heard thunder has experienced the effect of shock waves. A flash of lightning instantly heats the air to as much as 60,000°F (33,320°C). When air is heated, it expands. When it is heated to such a high temperature so quickly, air expands explosively and forms a shock wave. The shock wave rushes away from the lightning faster than the speed of sound. Within a few feet, it has slowed down and become an ordinary sound wave, which we hear as thunder. Similarly, the sharp crack of a whip is produced when the tip of the whip goes faster than the speed of sound and sets off a shock wave.

Aircraft

When an aircraft flies through the air, it pushes the air in front of it out of its way, which causes disturbances in the air. Pressure waves travel away in all directions. The fastest they can move is the speed of sound. When the aircraft goes faster than the speed of sound, the pressure waves ahead of it cannot escape fast enough. They pile up together in front of the aircraft and produce a sudden jump in pressure—a shock wave. This shockwave spreads out from the aircraft's nose in the same way that a wave forms in front of a ship's bow. Another shock wave spreads out from the aircraft's tail as air rushes into the hole left behind by the plane, like the wake that trails behind a ship. Other parts of a plane, such as the wings and

◑ The Apollo 8 capsule was photographed during its reentry in 1969. The shockwave formed during reentry helps protect a spacecraft from the intense heat of a high-speed descent into the atmosphere.

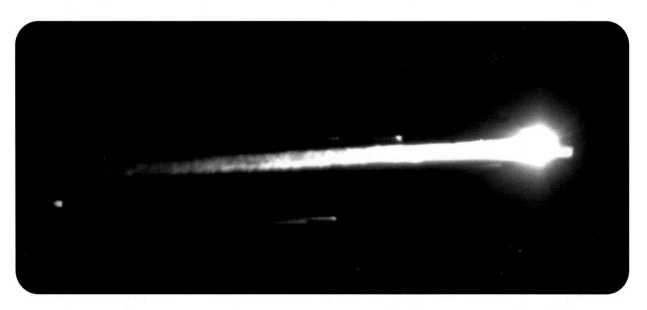

cockpit, produce more shock waves, but the nose and tail shock waves are the biggest.

Spacecraft

Shock waves also affect spacecraft returning to Earth. When a spacecraft plunges back into Earth's atmosphere from space, it compresses the air in front of it, and a shock wave is formed. When air is compressed, it heats up. The shock wave created by a spacecraft reentering the atmosphere at high speed is very hot. The Mercury, Gemini, and Apollo manned space capsules, were designed to reenter the atmosphere blunt end first. This blunt shape helped to push the super-hot shock wave forward, away from the capsule. A heat shield protected the astronauts and spacecraft from the heat that remained.

Today, the Russian Soyuz capsules use this method, and so does the Space Shuttle. The Space Shuttle reenters the atmosphere with its nose tipped up and its large underside facing the direction of flight. This deflects most of the heated shock wave around the spacecraft.

As the Space Shuttle descends lower in the atmosphere, nose and tail shock

SHOCK DIAMONDS

Sometimes, a line of bright spots called shock diamonds appears in the jet of hot gas from a jet engine or rocket. When the supersonic jet of gas from an engine or rocket slams into the air, the gas is squashed, forming a diamond-shape shock wave. The shock diamond is hotter than the surrounding gas, so unburned fuel from the engine is ignited, making the diamond glow. The jet expands and then it is squashed again, forming another glowing diamond, and another, and another.

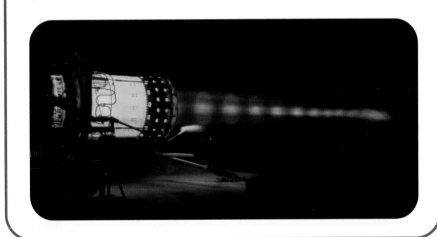

waves form on it. They spread out and reach the ground, where people hear two sonic booms. All supersonic aircraft form two shock waves. Unlike the Space Shuttle, however, most of these aircraft are so small, and the shock waves they create are so close together that only one bang is audible.

SEE ALSO:
- Air and Atmosphere • Pressure
- Supersonic Flight

Sikorsky, Igor

Date of birth: May 2, 1889.
Place of birth: Kiev, Ukraine.
Died: October 26, 1972.
Major contributions: Built the first successful single-rotor helicopter; built the first multi-engine airplane; built large aircraft used in early transoceanic passenger flights.
Award and Honors: Cross of St. Vladimir; Sylvanus Albert Reed Award; Presidential Certificate of Merit; Daniel Guggenheim Medal; Elmer A. Sperry Award; National Defense Award; National Medal of Science; Wright Brothers Memorial Trophy; Hawkes Memorial Trophy; Potts Medal; elected to National Aviation Hall of Fame; many honorary degrees.

Igor Sikorsky was a pioneer in the field of helicopters.

Igor Sikorsky was born in Kiev, Ukraine. He gained from his mother an interest in the work of artist and inventor Leonardo da Vinci, in particular his drawings of flying machines. At age twelve, Sikorsky made a model helicopter powered by rubber bands and made it rise into the air.

Two years later, in 1903, Sikorsky became a student at Russia's Naval Academy in the city of St. Petersburg. He became more interested in engineering and left to pursue his studies. After two years of schooling in Kiev, Sikorsky dropped out to focus on experimental work of his own. That year, 1908, he also developed a passion for flight. After seeing photographs of the Wright brothers' successful flights, Sikorsky later recalled, "I decided to change my life's work. I would study aviation."

Early Experiments

In the spring of 1909, Sikorsky built his first real helicopter. However, the machine would not fly. Another version failed to fly the following year, and Sikorsky decided to abandon the effort. As he later explained, "I had learned enough to recognize that with the existing state of the art, engines, materials, and—most of all—the shortage of money and lack of experience . . . I would not be able to produce a successful helicopter at that time."

Sikorsky turned his attention to designing airplanes, producing several models and flying them himself. In 1911, he earned an international pilot's license, becoming just the sixty-fourth person in the world to have one. That year, Sikorsky's S-5 plane set records by carrying three people more than 30 miles (48 kilometers) at 70 miles per hour (about 113 kilometers per hour). Another of his planes won an award at an air show the next year and took first prize in a competition held by the Russian armed forces. This success earned Sikorsky a job with a Russian company, where he worked on manufacturing airplanes.

In 1913, Sikorsky produced a new design he called the Grand. This large airplane was powered by four engines—the first flying machine to have more than one. It also was the first to have the pilot and passenger areas fully enclosed. The Russian army used several dozen aircraft of this design as bombers during World War I (1914–1918).

Russian participation in the war ended when communists took control of the nation's government and pulled the country's troops out of the conflict. Sikorsky left his homeland in 1919 and eventually reached the United States.

Success in the United States

Sikorsky struggled during his first few years in the United States. In 1923, with the assistance of several other Russian exiles, he formed Sikorsky Aero Engineering Company to build airplanes.

🎧 Sikorsky (seen here in the cockpit of a U.S. Coast Guard HNS-1 Hoverfly) saw the helicopter as a useful machine. As early as 1944, when this photo was taken, Sikorsky helicopters were used for rescue missions.

Within a few years, Sikorsky was building successful aircraft again. His S-29 carried fourteen passengers. With two engines, it could reach a speed of 115 miles per hour (185 kilometers per hour). Sikorsky gave it an all-metal body.

Sikorsky's company also produced the S-38, a ten-seater that could land on water. Pan American Airways bought several of these planes as the airline began to build its network, providing an air service to South America. Sikorsky based his company at Stratford, Connecticut. He became an U.S. citizen when he was naturalized in 1928.

Sikorsky ran into problems in 1929. He had built and sold several expensive planes, called "flying yachts," to wealthy businessmen. The planes were not all paid for when, in October 1929, the stock market in New York City crashed.

🎧 *Marine One* is the name of the helicopter that flies the U.S. president. Traditionally, *Marine One* is a Sikorsky. The 2005 model, shown here carrying President George W. Bush to the White House, was a Sikorsky VH–3D Sea King.

"I have never been in the air in a machine that was as pleasant to fly as the helicopter. It is a dream to feel the machine lift you gently up in the air, float smoothly over one spot for indefinite periods, move up or down under good control, as well as move not only forward or backward but in any direction."

Igor Sikorsky

Most of Sikorsky's customers lost their fortunes when the stocks they owned plunged in value. As a result, many of the buyers did not make their promised payments. Losing money, Sikorsky sold out to the United Aircraft Corporation. Sikorsky continued to produce planes for Sikorsky Aircraft, which has remained part of what is now United Technologies Corporation (UTC).

Sikorsky's next achievement was one of his most impressive. In 1931, the company launched the S-40, or the American Clipper. This large flying boat carried four engines. Pan American bought the planes and, by the late 1930s, was using Clippers to provide air service across both the Pacific and Atlantic oceans.

Helicopter Pioneer

United Aircraft discontinued the Clippers, but the company funded Sikorsky's effort to achieve his long-held dream of building a helicopter. This time, that effort succeeded, with the help of new, lightweight materials and a staff of engineers.

On September 14, 1939, Sikorsky climbed into the first helicopter model, the VS-300. He always insisted on taking the first flight of any completely new design. The helicopter worked—it rose vertically, hovered, and returned to land. The flying machine had a single rotor with three blades driven by a 75-horsepower engine. Sikorsky's helicopter was not the first to reach the air, but it was the first successful flight of a helicopter with a single rotor. Because most helicopters follow that design, Sikorsky is considered the leading pioneer of the helicopter industry.

The design needed improvement, however. Sikorsky tried another version with two small rotor blades in the rear. On May 13, 1940, that machine rose into the air, but was difficult to move forward. The next model had just one smaller rotor blade in the rear. This version flew smoothly and on May 6, 1941, it set a record by staying aloft for more than an hour.

Helicopters Take Off

The U.S. military signed a contract with Sikorsky Aircraft in 1943 to buy helicopters. The company began producing the R-4, the first mass-produced helicopter. The machines had little impact during World War II, but by the Korean War (1950–1953), they were in constant use.

During the 1950s, Sikorsky opened a new plant dedicated to making helicopters. Along with making the flying machines, he helped promote their use. A New York company used helicopters

to carry passengers between the city's different airports. The aircraft also were used to rescue people caught in disasters or to bring supplies to places difficult to reach in other ways. In 1950, the Collier Trophy was awarded to the entire helicopter industry. Sikorsky, who had pioneered the field, had the honor of accepting the award.

Sikorsky retired in 1957. He remained active in aviation and was elected to the Aviation Hall of Fame in 1968. On October 25, 1972, he was still working at his desk. He died the next day.

Sikorsky helicopters such as these are still produced today.

SEE ALSO:
• Aircraft Design • Da Vinci, Leonardo • Flying Boat and Seaplane • Helicopter • World War I

Skydiving

Skydiving is a sport in which a parachutist falls through the air, performing various maneuvers, before opening his or her parachute. Free-falling in this way is the closest that a human comes to flying like a bird.

Parachutists in the early 1900s opened their parachutes within a second or two of jumping. As parachuting developed, however, bolder jumpers delayed pulling the ripcord. They experienced free fall for a few seconds before the parachute opened and slowed their descent. This was the start of skydiving.

In the 1930s, parachuting began to develop as an international sport. The first world championships were held in 1951, and today there are competitions for accuracy, style, and team jumping.

⏻ Fifteen skydivers hold hands at 9,000 feet (2,740 meters) as another prepares to join them, having just jumped from a helicopter above.

Many people enjoy the thrill of skydiving as a recreational sport.

Flying Like a Bird

The first parachutists usually fell "like a sack being hurled out of a window," in the words of Leo Valentin, a parachute jumper. (Valentin wore suits with batlike wings in an attempt to imitate bird flight. Wing suits often broke in midair, and Valentin was killed in an accident in 1956.) Parachutists soon found, however, that by adopting a box position as they fell—stomach-down, with arms and legs bent slightly backward—they could soar like a bird rather than drop like a sack until their parachute opened.

Modern skydivers are taught to fall in the box position, although experienced divers also adopt other positions. The spread-eagle body acts like a wing, so a skydiver can fly around, and teams can formate (group together). Groups of as many as 282 divers have achieved formation in free fall.

486

An amazing demonstration of free fall maneuvering took place in 1987, in the sky above Arizona, when skydiver Gregory Robertson saved the life of fellow parachutist Debbie Williams. She was knocked unconscious after colliding with another jumper, and she fell for 6,000 feet (1,830 meters). Robertson dived alongside her and opened her parachute at 3,500 feet (1,070 meters) above the ground. Only then did he open his own parachute.

What Skydivers Do

Skydivers usually jump from an airplane, although they also may leap from helicopters and balloons at various heights. A typical jump height is between 10,000 and 13,000 feet (3,050 and 3,960 meters). On August 16, 1960, Captain Joseph Kittinger of the U.S. Air Force stepped out of a balloon over New Mexico at a height of 102,200 feet (31,150 meters). He experienced a record-breaking free fall, skydiving for 4 minutes and 38 seconds before he opened his parachute at 17,500 feet (5,330 meters).

In free fall, a skydiver does not feel a falling sensation, even though speed through the air may reach 120 miles per hour (193 kilometers per hour). Skydivers do not experience the discomfort of acceleration. During the descent, they may perform such maneuvers as turns, front and back loops, barrel rolls, and joining up with other skydivers in formations. There is a deceleration sensation when the parachute opens,

Members of a women's skydiving team descend after opening their parafoils. Use of the parafoil has increased distance and maneuvering ability for skydivers.

slowing the descent to about 12 miles per hour (19 kilometers per hour).

The Parachutes

To end a dive, the skydiver first pulls out a pilot parachute. Measuring about 3 feet (0.9 meters), the pilot parachute is stored in a pocket on the harness, or rig. As this small parachute inflates in the wind, a cord known as the bridle operates the release mechanism for the main parachute and its lines. A rectangle of fabric (the slider) separates the

THE PARAFOIL

The invention in the 1960s of the wing parachute, or parafoil, revolutionized sports parachuting. The modern parachute, known as a ram air wing, flies almost like a paraglider. The canopy contains seven to nine panels, or cells, open at the front so that air can enter. The parachutist can alter the amount of air inside the canopy by twisting a handgrip. By tugging on the steering lines, the parachutist can make turns and steer toward a landing site.

parachute lines into four groups and works its way down until the canopy is fully open and the slider is just above the skydiver's head.

A skydiver carries two full-size parachutes, one of which is kept in reserve. The minimum safe height for opening a parachute is around 2,000 feet (610 meters). This height gives sufficient time for the skydiver to open the reserve parachute if the main parachute fails to function. Many skydivers carry an automatic activation device (AAD), which opens the reserve parachute at a safe altitude if the skydiver fails to open the main parachute. Skydivers always carry an altimeter so they know at what height to open the parachute. If the main parachute malfunctions after it has been opened, the skydiver uses one handle to discard it and pulls another handle (on the parachutist's chest) to open the reserve parachute.

Rectangular canopies are better than traditional round canopies for competition jumping as they are much easier to steer. A rectangular canopy will not collapse should another parachutist fly beneath it and take its air. Using these modern canopies, skydivers can fly in stacks, one above the other. An experienced skydiver with a modern parachute can fly cross-country for as much as 10 miles (16 kilometers), reaching speeds of up to 30 miles per hour (48 kilometers per hour) over the ground. He or she can perform dramatic maneuvers such as the swoop: a fast, downward approach before leveling off just above the ground.

Learning to Skydive

Free-fall parachuting is best taught by an experienced instructor. After ground instruction, the first jumps are often made in tandem (instructor and student together). The tandem jump also offers a way for elderly or physically disabled people to skydive.

Most people experience fear when beginning to skydive. Training helps the beginner develop confidence before going on to practice advanced techniques with fellow skydivers. The U.S. Air Force Academy, for example, trains cadets for more than 33 hours on the ground before their first jump. Advanced training consists of more than 150 free fall jumps, progressing from a large (and therefore slow) canopy to a high-performance display parachute with a vertical descent speed of 2 to 16 feet per second (0.6 to 5 meters per second). Top students may go on to jump with the Wings of Blue display team.

Safety and Regulation

Although accidents do happen, skydiving and parachute sports have a good safety record. Accidents are most common when people jump in poor weather conditions, such as unpredictable winds. Jumping from buildings, cliffs, or other high structures (known as BASE jumping) is especially dangerous. Because the modern parachute can be steered, there is little chance of the parachutist landing accidentally in a lake or a tree, as was often the fate of parachutists in the past.

Drop zones in the United States and Canada are required to have an experienced person who acts as a safety officer. In most countries, skydivers are required to carry a reserve parachute that has been packed and inspected by a certified parachute rigger. In the United States, certification is provided by the Federal Aviation Administration (FAA).

Many countries have national parachuting associations, affiliated to the Fédération Aéronautique Internationale (FAI). In the United States, skydiving permits and ratings are issued by the United States Parachute Association.

SEE ALSO:
• Glider • Hang Glider • Parachute

EXTREME SKYDIVING

Specialized forms of skydiving and parachuting include:
• Accuracy landing: Aiming to land on or very near a drop zone target.
• Blade running: Like slalom skiing with a parachute.
• Formation skydiving: Also called relative work (RW).
• Paraskiing: Landing on a snowy mountain on skis.
• Skysurfing: Landing with a surfboard strapped to the feet.
• Stuff jumping: Jumping with an object, such as a bicycle, which is ridden through the air before the skydiver lets go and opens the parachute.

A skydiver BASE jumps from one of the world's tallest buildings in Shanghai, China. The "BASE" in BASE jumping is an acronym that stands for building, antenna, span, and Earth.

Skyjacking

Skyjacking is the illegal seizure of an airplane. It is a crime similar to hijacking a truck or taking over a ship at sea. Skyjackers may demand that the plane be flown to a destination of their choice or demand a ransom for the release of passengers. They may use the airplane as a weapon of destruction.

How Skyjacking Began

The first recorded skyjacking was in 1931 in Peru. Rebel soldiers forced two American pilots to fly a plane over the city of Lima to drop propaganda leaflets. The first skyjack in the United States

A 1985 skyjacking in Beirut, Lebanon, ended when the hijackers released the hostages and then blew up the Boeing 727 they had hijacked. One of the hijackers is seen here on the runway.

DISAPPEARING AIR PIRATE

Probably the most famous criminal skyjacking in the United States happened in 1971. A man known as Dan or D. B. Cooper took over a Northwest Orient Airlines Boeing 727. After forcing the plane to land, he demanded $200,000 as ransom for the release of the passengers. When he had received the money and four parachutes (one each for himself and the three remaining crew members), Cooper ordered the airplane to take off again. He parachuted from the rear of the Boeing 727 over the Cascade Mountains of the northwestern United States and was never seen again. There have been many suspects, but no certain culprit has ever been found.

took place in 1961, when a passenger on a commercial flight from Miami to Key West, Florida, ordered the pilot to divert the aircraft to Communist-ruled Cuba. With lax security at airports in the 1960s, it was relatively easy for a terrorist to smuggle a gun onto an airliner to threaten the pilots and passengers. Skyjackers often had political motives. Taking over an airliner ensured publicity for their cause. Less often, a skyjacker was a criminal who hoped to extort money by air piracy.

In the 1970s, when tension between Israel and its Arab neighbors was high, terrorists based in the Middle East made several attacks on airliners. Their usual practice was to seize an airliner in flight, force the pilot to land, and then broadcast their demands by radio. Skyjackers held passengers inside the airplane as hostages, hoping to bargain for the release from jail of fellow terrorists or other prisoners.

In September 1970, a spectacular skyjacking took place in the Middle East, when Palestinian terrorists seized three airliners simultaneously. All three aircraft were landed in the Jordan desert at Dawson's Field (a former British air force base) and then blown up after most of the hostages had been released. An attack on a fourth airplane was foiled by Israeli security guards. A fifth airplane was hijacked three days later.

🎧 U.S. military personnel train constantly to prepare for skyjackings and other terrorist acts. This photograph was taken during a U.S. Air Force skyjacking exercise.

Tightening Security

Such attacks highlighted a new threat to air travel and the need for tighter security at airports. The Hague Convention of 1970, an international agreement signed by more than 130 nations, was drawn up to combat skyjacking and stop terrorists from escaping to "friendly" countries.

From 1973, the Federal Aviation Administration in the United States has required all airlines to screen passengers and their baggage to prevent people from carrying weapons—or objects that might be used as weapons—onto flights. Airports tightened security procedures, especially at check-in and during baggage handling. Armed guards, called sky marshals, traveled on some flights.

When an airplane is hijacked, military craft may escort it to a landing field agreed to by the authorities. After it has landed, troops and police will surround the hijacked aircraft while skilled negotiators try to talk the skyjackers into releasing the hostages and surrendering. Armed assault also may be used. In 1976, following the hijacking of an Air France flight, Israeli commandos flew in to attack the Palestinian hijackers, who were holed up at Entebbe Airport in Uganda, Africa. The commandos rescued more than 100 hostages. In most hijacking situations, however, airport and law enforcement agencies usually try to avoid a gun battle, which risks injuring or even killing innocent hostages.

Skyjacking incidents decreased in the United States during the 1980s and 1990s, but violent terrorist incidents continued to take place in other parts of the world. Some of these skyjackings resulted in airplanes crashing. In 1996, a stolen Ethiopian airliner crashed into the Indian Ocean. About fifty passengers

managed to survive, but the crash killed 125 of the people onboard.

September 11, 2001

The worst skyjacking in history happened in the United States in 2001. On the morning of September 11, nineteen terrorists hijacked four U.S. airliners: American Airlines Flights 11 and 77 and United Airlines Flights 93 and 175. Flights 11 and 175 had taken off from Boston, Massachusetts; Flight 77 left from Dulles Airport in Washington, D.C.; and Flight 93 departed from Newark, New Jersey. The first three airplanes were on early-morning flights to Los Angeles, California; the fourth was heading for San Francisco, California.

Two of the hijacked planes (Flight 11 and Flight 175) were deliberately flown into the twin towers of the World Trade Center in New York City. Flight 11 hit the North Tower just before 8:45 A.M. Flight 175 hit the South Tower at 9:03 A.M. Both 110-story structures became fiery infernos, pouring black smoke into a blue sky, before collapsing to the ground. At 9:40 A.M., Flight 77 was flown into the side of the Pentagon in Washington, D.C.

On Flight 93, the hijackers—who had smuggled knives onboard—had locked themselves in the cockpit and headed the plane toward Washington, D.C. Flight 93's passengers learned from cell

↻ Terrorists flew an airplane into the Pentagon building in Washington, D.C., as part of their skyjacking operation on September 11, 2001.

phone calls to friends and relatives what had happened to the other three planes. They decided to attack the hijackers. The plane went out of control and crashed near Shanksville, Pennsylvania.

Every person onboard the four hijacked airplanes, including the hijackers, was killed. Furthermore, many more people were killed on the ground in New York City and Washington, D.C. The total death count was 2,752 at the World Trade Center, 189 at the Pentagon, and 44 in Pennsylvania.

After 9/11

Responsibility for the attacks was leveled at al-Qaeda, a secretive Islamist terrorist organization led by Osama bin Laden. U.S. President George W. Bush announced a "war on terror," and U.S. warplanes were ordered to shoot down any hijacked airliner that might pose a danger. No-fly zones were enforced.

Some of the 9/11 terrorists had been living in the United States and had even taken flying lessons there. The 9/11 attacks led to a review of the nation's security. Stricter antihijacking regulations were introduced to prevent explosives or weapons from being taken onto airplanes. Air marshals disguised as passengers traveled on flights, ready to disarm potential skyjackers. Within a few weeks, President Bush had signed a new law, the Anti-Terrorism Act, giving the U.S. government increased powers.

Other suicide attacks were foiled. Later in 2001, for example, law enforcement agents seized al-Qaeda terrorist

⬆ In New York City, two blue beams of light shine into the night sky in the place where the twin towers of the World Trade Center once stood.

Richard Reid (a British citizen), who had been planning to blow up a U.S. airliner with a bomb hidden in his shoe.

Today, passenger and baggage screening systems are provided by the Transportation Security Administration (TSA), part of the Department of Homeland Security. Under new secure flight arrangements, airlines and security services exchange information to identify all persons buying airline tickets, checking identities against those of known terrorists. Counterterrorist intelligence in the United States is spearheaded by the National Counterterrorism Center, which took over the State Department's responsibility in that area.

SEE ALSO:
• Airport • Pilot

Sound Wave

Sound is a form of energy that travels through the air as a series of pressure waves. The waves' effect on the ear produces the sensation of hearing.

Anything that vibrates produces sound waves that spread out through the air. When something vibrates, it pushes against the air next to it and squashes the air. This pressure creates a pulse that travels away through the air. After pushing against the air, the object pulls back again before the next push. When it pulls back, it creates a region of low pressure in the air. These high-pressure pulses are called compressions, and the low-pressure regions between them are known as rarefactions. The sound waves they create are called longitudinal waves.

The distance between one compression and the next is a sound wavelength. The number of compressions that pass any point in a second is the sound's frequency. A high-pitched whistling sound has a high frequency, or a lot of pulses per second. A deep rumbling sound has a low frequency. Frequency and wavelength are linked. As frequency increases, wavelengths get smaller.

Sound needs something to travel through. It can travel through liquids and solid materials as well as air and other gases, but it cannot travel through space without air. The speed of sound in air depends on how fast vibrations spread from molecule to molecule, and this depends on the temperature of the air. Sound travels faster in warm air than in cold air. On an average day, with an air temperature of about 59°F (15°C), the speed of sound in air is about 760 miles per hour (1,220 kilometers per hour).

Vibrations produce sound, but sound itself also can make things vibrate. We hear because sound waves enter our ears and make our eardrums vibrate. Every machine, and every part of a machine, has a resonant frequency at which it

TECH TALK

MEASURING SOUND

The frequency of a sound is measured in hertz (abbreviated as Hz). One hertz is equal to one compression per second. The human ear can hear sounds ranging in frequency from about 20 hertz to 20,000 hertz. Sounds below 20 hertz are called infrasound, and sounds above 20,000 hertz are called ultrasound.

Sound intensity is measured in units called decibels (abbreviated as dB). The quietest sound that can be heard has an intensity of 0 decibels. A sound ten times more intense is 10 decibels. A sound 100 times more intense is 20 decibels. A sound 1,000 times more intense is 30 decibels. A whisper is about 20 decibels, while a jet engine 100 feet (30 meters) away is about 150 decibels.

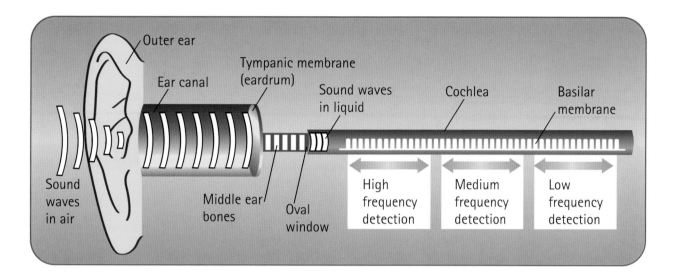

Outer ear

Tympanic membrane
(eardrum)

Ear canal

Sound waves
in liquid

Cochlea

Basilar
membrane

Sound
waves
in air

Middle ear
bones

Oval
window

High frequency detection	Medium frequency detection	Low frequency detection

vibrates very strongly. Engineers try to ensure that aircraft and their engines do not vibrate at their resonant frequency. This is because strong vibrations can cause damage, shake things loose, and make an aircraft very noisy.

Vibrating rotors and propellers can cause a lot of noise inside helicopters and propeller planes. The noise makes it difficult for pilots to hear messages in their radio headsets. To counteract this, pilots can wear headsets that remove unwanted sound by making more sound. The system listens to the background noise and instantly makes a copy of it with the compressions and rarefactions reversed. The compressions of one sound fill up the rarefactions of the other sound, and thus the two sounds cancel each other out.

SEE ALSO:
• Air and Atmosphere • Communication • Energy • Pressure

⬆ This diagram shows how sound waves enter a human ear and reverberate inside the eardrum so we can hear sounds.

⬆ The Space Shuttle has to be protected from the roar of its own engines when it is launched. to prevent damage to the spacecraft. Just before the engines fire, 300,000 gallons (1,135,500 liters) of water start pouring onto the launch pad, as shown in this photo taken during a system test. The water absorbs the sound and stops it from bouncing back to the spacecraft.

Spaceflight

Spaceflight means traveling in space. Manned spacecraft of different kinds, from the space capsules of the 1960s and 1970s to the Space Shuttle, have carried people on spaceflights. Spacecraft called satellites are in flight as they orbit Earth. Space probes have visited planets such as Mars, Venus, and Jupiter, and they have even flown beyond the solar system.

WHERE DOES SPACE START?

There are several definitions for where space begins. The atmosphere, a blanket of air around Earth, gradually thins as altitude (distance from Earth's surface) increases. Earth's atmosphere fades away almost completely at the top of the thermosphere, which is about 400 miles (640 kilometers) from Earth's surface. In terms of spaceflight, however, people define space as beginning much closer to home, either 50 miles (80 kilometers) or 62 miles (100 kilometers) above Earth's surface. Space scientists and engineers refer to "entry interface" as being just over 75 miles (120 kilometers) above Earth's surface. This is the point at which the air becomes thick enough to begin heating up a spacecraft as it returns from space.

⊙ The invention of powerful rockets, along with other significant developments in technology, has enabled human beings to launch themselves into space.

Launch and Reentry

Spaceflight became possible with the development of rockets that had sufficient power to break free of Earth's gravity. To break free, a rocket must reach escape velocity, which is just over 25,000 miles per hour (40,200 kilometers per hour). Takeoff and reentry are the two most dangerous times in a spaceflight. Before the 1950s, some scientists argued that the human body could not survive the stresses of a space launch. The earliest flights by astronauts proved such views wrong. Astronauts have flown faster than any humans have before and have returned back to Earth unharmed.

Conventional rocket motors work both in air (for takeoff) and in space. Most rockets used to launch spacecraft are multistage vehicles propelled by chemical fuel burned in liquid or solid form. The propellants must include oxygen, or the fuel will not burn, because there is no air in space.

Booster rockets provide extra thrust during takeoff. In a multistage rocket, boosters and lower stages separate and fall away as soon as their fuel is burned up. Only the topmost stage reaches space. The load a rocket lifts into space is called its payload—this could be a satellite, a manned spacecraft, or a robot space probe.

A rocket is streamlined for efficient, controlled, high-speed flight through the air. A spacecraft designed to return to Earth, like the Space Shuttle, also is streamlined, but it has wings. The wings help the Space Shuttle land like a conventional airplane after it has reentered Earth's atmosphere.

Returning to Earth from space is potentially as dangerous as leaving it. A spacecraft must decelerate (slow down), using braking rockets, and approach at a precise angle so it does not hit the Earth's atmosphere too fast. Reentry is accompanied by a rapid rise in temperature. Air gets trapped in front of the spacecraft, which is moving so fast the air cannot escape. Compression (squeezing) of the air raises the temperature to more than 10,000°F (5,540°C). Spacecraft would burn up unless protected by a heat shield of tough, insulating material.

Traveling in Space

After being launched, a spacecraft leaving Earth for the Moon or Mars does not need to keep burning fuel. Spacecraft can make use of other methods of propulsion, such as solar sails or ion engines. Once on course and free of Earth's gravity pull, the engines can be switched off to save fuel as the spacecraft coasts through space. This is how the Apollo astronauts traveled to the Moon, a trip that took two-and-a-half days. They fired their engines only to slow down the spacecraft and during their return to Earth.

The Apollo 10 crew sped home at 24,790 miles per hour (39,890 kilometers per hour) on their return from the Moon in 1969. The astronauts' capsule splashed down safely in the Pacific Ocean.

🎧 *Helios A* and *Helios B* were space probes sent in the 1970s to orbit the Sun. They reached the highest speed of any spacecraft. A 1974 photograph shows *Helios A* on top of a launch vehicle.

Although there is no air in space, space is not empty. It contains dust, chunks of minerals, space junk, and streams of radiation flowing out at great speed from the Sun and from other stars.

Stretching into space around Earth is a magnetic field. This magnetism attracts electrically charged particles that form belts, or zones, of radiation. Named the Van Allen radiation belts, these radiation zones were unknown until the first U.S. satellite, *Explorer 1*, encountered them in 1958. The Van Allen belts were the first important scientific discovery made by a spacecraft.

The fastest spacecraft sent from Earth so far have been the solar probes *Helios A* (1974) and *Helios B* (1976). *Helios B* traveled about 150,000 miles per hour (241,350 kilometers per hour) as it orbited the Sun. Although spacecraft are the fastest vehicles ever flown by humans, they are snail-like in space terms, where the distances are unimaginably immense. The nearest star is 4.2 light years from Earth. So even if a future spacecraft could reach light speed of 186,000 miles per second (299,280 kilometers per second), it would take 4.2 years to get there.

To fly astronauts to Mars and back using existing spacecraft would take eighteen months. Keeping astronauts alive, healthy, and able to work during such a long mission poses great challenges to space science. Humans are not designed for an airless, weightless environment. A manned spacecraft must provide everything needed for human life support—air, water, food, fuel, energy, waste disposal, and exercise. Long periods of spaceflight weaken the body's muscles. So great are the challenges that some scientists believe that robot probes are safer, more cost effective, and better suited for exploring the planets.

Pioneers of Spaceflight

Spaceflight began in the middle of the twentieth century, but scientists and writers had imagined the possibility long before. In the seventeenth century, physicist Sir Isaac Newton set out laws of motion that determine the way in which objects move through space. Fantasies about space travel came from science-fiction writers such as Jules Verne. In his 1865 book *From the Earth to the Moon*, Verne wrote of people flying to the Moon in a capsule fired from a huge cannon. In 1898, H. G. Wells imagined Martian spacecraft invading Earth in *The War of the Worlds*. At this time, people could only study the Moon and Mars by peering through optical telescopes, and there were many fanciful notions about alien life-forms on distant worlds.

Russian teacher Konstantin Tsiolkovsky (1857–1935) figured out the mathematical principles of spaceflight by rockets. In 1923, Hermann Oberth (1894–1989) wrote *The Rocket into Planetary Space*, a book that predicted spaceflight.

JAMES A. VAN ALLEN (1914–2006)

James Van Allen was born in Mount Pleasant, Iowa, and studied physics at the University of Iowa. During World War II, he designed parts for anti-aircraft guns and then served with the U.S. Navy in the Pacific. In 1951, Van Allen became head of the Physics Department at the University of Iowa, where he taught for more than thirty years. A renowned astrophysicist, he was one of the first American scientists to propose launching satellites. Using equipment installed by Van Allen, the first U.S. satellite *Explorer 1* (January 1958) detected two belts of electrically charged particles orbiting Earth. They were named after Van Allen, who later discovered similar radiation belts around the planet Saturn. Professor Van Allen received many awards for his work, including the Gold Medal of the U.K. Royal Astronomical Society; the National Medal of Science, 1987; the Vannevar Bush Award, 1991; and the National Air and Space Museum Trophy, 2006.

James Van Allen (center) stands with William H. Pickering (left), former director of the Jet Propulsion Laboratory, and Dr. Wernher von Braun, leader of the team that built the rocket that launched *Explorer 1*. The three men display a model of the satellite after its successful launch in 1958.

🎧 *Voyager 1* took photographs of Jupiter and its four planet-size moons, and the images were assembled to form this composite photo. Unmanned spaceflights into deep space are expanding human knowledge of the universe.

Johannes Winkler Oberth (1897–1947), along with other German enthusiasts, formed the Society for Space Travel. One of its members was Wernher von Braun (1912–1977), who helped design the V-2 rocket of World War II and later worked on the U.S. space program. In 1926, American Robert H. Goddard (1882–1945) launched the world's first liquid-fuel rocket.

The American Interplanetary Society was founded in 1930 by G. Edward Pendray, David Lasser, Laurence Manning, and others. In 1934, it became known as the American Rocket Society.

In 1963, it became part of the American Institute of Aeronautics and Astronautics. The American Rocket Society and the British Interplanetary Society both helped stimulate public interest in space-flight and encouraged test flights of rockets at a time when governments had little interest in spaceflight.

The Space Age

In 1947, a WAC-Corporal rocket reached a height of 244 miles (390 kilometers) above the White Sands testing ground in New Mexico. This flight opened the door to space—and encouraged government interest—because it showed that rocket technology could now launch satellites. Ten years later, the Russians launched *Sputnik 1*, the first artificial satellite. So began the space race between the United States and the Soviet Union. Key figures in the two rival space programs were Robert Gilruth (1913–2000), appointed to direct the U.S. Mercury astronaut program in 1958, and Sergei Korolyov (1906–1966), who led the Soviet design team behind the Sputnik program.

Spaceflight was front-page news through the 1960s as the United States and the Soviet Union competed to send astronauts into orbit and probes to the planets. Public interest in space reached a peak during the Apollo Moon landings (1969–1972).

Since the 1970s, spaceflight has developed as a multinational scientific activity and a commercial business.

LIFE BEYOND EARTH

The Planetary Society was founded in 1980 by Carl Sagan, Bruce Murray, and Louis Friedman. Its aim is to "inspire and involve the world's public in space exploration" and to search for extraterrestrial life (life beyond Earth). A worldwide program known as SETI (Searching for Extra-Terrestrial Intelligence) aims to collect evidence of life on other worlds through detecting radio signals and other forms of transmissions. Currently, Earth is the only planet known to support life, although scientists speculate that life could exist on Earth-type planets orbiting other stars.

Most satellite launches, for example, are intended for communications and entertainment, such as TV broadcasting. Spaceflight has become almost routine, although tragedies—such as the losses of two U.S. Space Shuttles, *Challenger* in 1986 and *Columbia* in 2003—reminded people just how dangerous space can be. Robot space probes have made astonishing voyages, not only to planets but far beyond our solar system. In 1990, NASA's *Voyager 1* probe took a photograph of Earth from a distance of 4 billion miles (6.5 billion kilometers) more than twelve years after it set out on a voyage through space that may well last hundreds of years.

Since the end of the Apollo program in the early 1970s, the most significant manned spacecraft has been the U.S. Space Shuttle. First launched in 1981, the Space Shuttle flies regularly into orbit, delivering supplies to the International Space Station.

The future of spaceflight will probably see a return of astronauts to the Moon and possibly a manned exploration trip to Mars. A lunar base might be in existence before the middle of the twenty-first century, and the discovery of water on Mars could even make a Martian colony a possibility.

SEE ALSO:
• Apollo Program • Astronaut
• Future of Spaceflight • Rocket
• Satellite • Space Probe

Space Probe

A space probe is a robot spacecraft sent far into space. To *probe* means to investigate, and that is what space probes do. Scientists have sent space probes to the Moon, to the planets, and beyond the solar system into deep space. A probe equipped with sensors, cameras, computers, and a radio transmitter can—within a few hours—discover more about a distant planet than has been gathered in centuries of Earth-based astronomy.

A robot space probe needs no air, water, or fuel during its journey through space. Its power comes from tiny nuclear plants or from solar cells that convert sunlight to electrical energy. Once set on its course, a probe can continue to travel through space for years, sending images and data back to Earth.

Flyby probes pass close to their target—a planet, a comet, or an asteroid. Orbital probes go into orbit around a planet or the Sun. Landers descend through the atmosphere to the surface.

Leaving Earth

Most probes are launched by a multi-stage rocket from the ground. There are three ways to send spacecraft into space using a rocket: sounding trajectory, Earth orbit, and Earth escape.

Sounding rockets were often fired into space during the 1940s and 1950s and are still used today. A sounding rocket can be fired to an altitude of about 100 miles (160 kilometers), at a

HIGH-SPEED LAUNCH

The fastest space launches have all involved space probes. In 1972 NASA's *Pioneer 10* was launched toward Jupiter at 32,400 miles per hour (52,130 kilometers per hour). In 1990, the probe *Ulysses*, on a mission to study the Sun, reached 34,450 miles per hour (55,430 kilometers per hour) during launch. *New Horizons*, launched in 2006 toward Pluto, was boosted to 35,800 miles per hour (57,600 kilometers per hour), as it left Earth's orbit for deep space.

↑ *Pioneer 10*, launched in 1972, was the first spacecraft to fly through the asteroid belt that lies between Mars and Jupiter, into the outer regions of the solar system.

⮩ A NASA sounding rocket is fired in 1988. Sounding rockets only reach the fringes of space, but they offer an inexpensive way of gathering data.

maximum speed of about 5,000 miles per hour (8,050 kilometers per hour). After its engine burns out, the rocket begins its descent back to Earth. Scientific instruments in the nose of the sounding rocket send information to the ground by telemetry (radio) or may be retrieved by parachute.

To enter Earth orbit, a rocket trajectory must be at an angle so that it flies parallel to Earth's surface. When its booster motors cut out, the topmost stage of the rocket must be going fast enough to enter orbit and not fall back to the ground under the pull of Earth's gravity.

To escape completely from Earth's gravity and become a planetary probe, a spacecraft must reach a velocity of around 25,000 miles per hour (40,200 kilometers per hour). It will then fly away from Earth, gradually slowing down. It may go into orbit around the Sun, or it may be attracted by the gravitational pull of a planet, such as Mars or Jupiter.

Staying on Course

The calculations involved in space navigation are complex. The launch base (Earth), the space probe, and the probe's target (a moon of one of the giant planets, perhaps) are all moving through space. Ground controllers must calculate launch speed and course precisely. If necessary, they make midcourse corrections by using computers to fire small rocket motors onboard the spacecraft. In this way, scientists can send a probe on voyages that will last for years.

When deciding on a launch date for a planetary probe, scientists choose a favorable "window," usually when the target planet is at its closest. Moving between planets in space seldom involves traveling in a straight line. Keeping a space probe on the right course requires smart computing and

🎧 A conceptual illustration by NASA shows how samples could be launched from the surface of Mars in a capsule that would bring them back to Earth as part of future missions to Mars.

Landing

When it reaches its destination, a probe may stay in orbit, radioing data and images back to Earth, or it may attempt to land a capsule on the surface. Landing on a planet many millions of miles away, under remote control, is always a challenge. The speeds of approach can be enormous. In 2003, the *Galileo* probe accelerated to 108,000 miles per hour (173,780 kilometers per hour) as it dived toward Jupiter.

Once on the target planet, a lander can use remote-controlled arms and scoops to collect samples of rock and soil. Its instruments analyze the samples and the gases in the atmosphere and measure temperature, pressure, and radiation levels. A few probes have released a small rover to explore the areas farther from the lander.

Some probes are sent to collect material and return it to Earth at the end of their mission. A reentry capsule drops down through Earth's atmosphere, by parachute, for recovery on the ground or in the air using the "air snatch" technique, by which an airplane scoops up the capsule before it hits the ground.

To the Moon

When the space age began in the 1950s, scientists were eager to expand their knowledge of the worlds beyond Earth, previously seen only through telescopes. After the launch of the first satellites by the United States and Soviet Union in 1957 and 1958, the world waited expectantly for the first

accurate gyroscopes on the spacecraft. The gyroscopes are used for inertial guidance to keep the space probe on course without reference to the Sun or stars. Instruments measure the slightest change in the spacecraft's acceleration so that computers can calculate any adjustment to the course. When planning a multiplanet mission, scientists may be able to send the probe on a "slingshot" trajectory. This takes the probe around one planet and then uses the planet's gravitational pull to accelerate it off to the next target. *Pioneer 11* did this in 1975, swinging around Jupiter onto a path that took it to Saturn.

rocket shot at the Moon. This came in January 1959, when the Soviet probe *Luna 1* flew within 3,700 miles (5,920 kilometers) of the Moon. Two months later, the United States sent its probe *Pioneer 4* to fly by the Moon. In September 1959, the Soviet *Luna 2* probe crashed onto the Moon. *Luna 3* flew around the Moon in October 1959 and took photographs of the far side, never before seen from Earth.

The First Planets

Venus was the first planet to be reached by a space probe. In 1962, the U.S. probe *Mariner 2* flew within 22,000 miles (35,400 kilometers) of Venus, but the Russians made the first remote-controlled landing in 1970, with their probe *Venera 7*. The U.S. *Magellan* spacecraft arrived at Venus in 1990. During a four-year stay, it sent back radar images of almost the entire planet surface.

In the 1960s and 1970s, U.S. Mariner probes investigated

the planet Mars as well as Venus and Mercury. *Mariner 9*, launched in 1971, went into orbit around Mars and sent back the first close-up pictures of the planet. In 1976, the United States landed *Viking 1* and *2* on Mars.

In November 1996, *Mars Global Surveyor* became the sixteenth space probe to fly by, orbit, or land on Mars. The following year, 1997, the *Pathfinder* lander made a touchdown on Mars and released a robot rover named *Sojourner*. The little solar-powered rover had a spectrometer to analyze the chemical composition of the Martian soil and a camera to send back pictures of the surface.

The Voyager Probes and *Galileo*

Much of what scientists now know about the four "gas giant" planets—

➲ The identical space probes *Voyager 1* and *2* continue to travel decades after they were launched. It is hoped that they will continue their journey farther into space. *Voyager 1* has already reached the outer edge of the solar system.

Jupiter, Saturn, Uranus, and Neptune—came from NASA's two Voyager probes, launched in 1977. *Voyager 1* flew past Jupiter and Saturn before leaving the

DISTANT VOYAGERS

Voyager 1 and *Voyager 2* each carry a gold disk showing the location of Earth within the Milky Way galaxy. The golden record also contain sounds and images chosen to portray the diversity of life on Earth. It is meant to communicate with any intelligent life-form that might collect one of the Voyagers. The Voyager spacecraft will take about 40,000 years to approach another star, however, and the probes are minute compared to the vastness of interstellar space. The chances of any alien life-form finding one of the probes is therefore remote.

solar system. *Voyager 2* journeyed on to visit Uranus in 1986 and Neptune in 1989. Jupiter has some of the wildest weather in the solar system, with winds up to 300 miles per hour (480 kilometers per hour). Jupiter also spins faster than any other planet. As a result, its day is less than 10 hours long.

In 1995 the *Galileo* probe orbited Jupiter and sent a small, cone-shaped lander plunging down into the atmosphere through clouds of ammonia ice crystals. The probe survived for an hour, sampling the hostile atmosphere, until it was destroyed.

The Cassini-Huygens Mission

The *Cassini* spacecraft launched to Saturn in 1997 reached the planet in 2004. In January 2005, the spacecraft released the *Huygens* probe to explore Saturn's moon, Titan. Parachutes slowed *Huygens's* final descent, and its cameras began taking pictures of Titan's surface from a height of 10 miles (16 kilometers). Finally, the probe landed on what looked like a shoreline—perhaps beside a lake of freezing liquid methane gas.

The *Huygens* probe continued to transmit data for 90 minutes, three times longer than scientists had hoped for. Signals from *Huygens* were transmitted to *Cassini* in orbit, and from *Cassini* back to Earth, where 45 minutes later they were picked up by large radio telescopes. The scientific instruments onboard the *Huygens* probe gave scientists much valuable data about Saturn's large and distant moon.

Return to Mars

In 2004, NASA returned to Mars with twin robot rovers named *Spirit* and *Opportunity*. During the landing, each rover was protected inside a large airbag with a parachute attached. After impact, the ball bounced over the Martian surface until it came to a halt. Then the airbag deflated and opened to release the robot rover. The rovers landed at separate locations. One of the mission aims was to look for water—a discovery that would make future landings on Mars by astronauts a more realistic prospect. Landing during the Martian afternoon, with Earth in full view, meant that the landers could signal at once to the waiting scientists to let them know that the landing had been successful. The signals were sent to Earth by way of the Deep Space Network, a series of antennae in California, Spain, and Australia. *Spirit* and *Opportunity* were intended to work for about 90 days, but they were still busy two years after they landed. They found evidence that Mars was, in its past, apparently a watery planet.

Probing the Future

NASA provides up-to-date information on its current space probe missions through its Web sites. The Jet Propulsion Laboratory provides updates on missions in progress and on future missions. NASA's Web sites also display the latest photos taken by space probes.

Robot probes will continue to play an important part in twenty-first-century

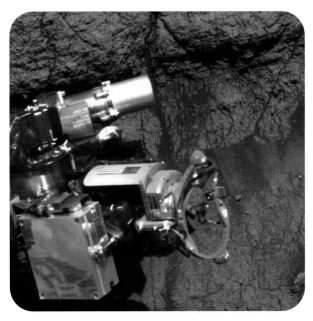

The Mars rover *Opportunity* took this photo of itself in 2004. It shows its rock abrasion tool after it ground into some Martian rock that covered it with red dust.

space exploration. Some already have set out on their long journeys. The *Messenger* probe left Earth in 2004 and will arrive at Mercury in 2011. The *New Horizons* probe, launched in 2006, should reach Pluto in 2015.

Space probes continue their journeys into infinity, long after they have ceased to communicate with Earth. Scientists calculate that by the year 34,593, *Pioneer 10* will have reached a star called Ross 248, 10.3 light years distant from Earth.

SEE ALSO:
- Gravity • Rocket • Satellite
- Spaceflight • Velocity

VOLUME GLOSSARY

Please see Volume 5 for a complete glossary for all volumes.

Aerobatics Form of aviation in which pilots fly in patterns and perform stunts, both singly and in formation.

Aeronautics Science of designing, making, and flying aircraft.

Air pressure Force of air pressing against something.

Airlock Airtight chamber that forms a passage between two spaces with differing air pressure.

Altimeter Instrument that measures altitude.

Antenna Device, such as a radio wire or a large dish, used to receive or transmit electromagnetic waves.

Artillery Weapons, including large guns and cannons, that fire shells, rockets, and missiles.

Astrophysics Science that combines physics and astronomy to study the formation, processes, and behavior of stars and space.

Attitude Angle of an aircraft in relation to the horizon or in relation to its surroundings.

Axis Line around which something rotates.

Barnstormer Stunt pilot of the early 1900s who flew in aerobatic exhibitions.

Beacon Radio transmitter that sends out signals to guide aircraft.

Booster rocket Another term for a launch vehicle or rocket booster.

Canard Small, movable wing placed toward the front of an aircraft, often positioned on an airplane's nose.

Control surface One of the surfaces (such as an aileron, elevator, or rudder) on an aircraft that are used to change direction.

Delta wing Airplane wing with a long, swept-back leading edge and a short, straight trailing edge.

Docking Joining one spacecraft to another in space.

Elevator Movable panel on an airplane's tailplane that controls the aircraft's pitch (movement about its lateral axis).

Elevon Tilting part at the back of an aircraft wing that performs the combined jobs of an aileron and an elevator.

Flight simulator Apparatus set up like an aircraft cockpit that can simulate flight conditions and is used for flight training or aviation research.

Frequency Number of cycles per second of sound, light, radio, or other waves.

Galaxy Huge clusters of stars, gas, and dust, such as the Milky Way that contain Earth's solar system and many others.

Gravity Attraction of objects to the center of Earth or to another planet or body.

Guided missile Missile that uses a guidance system, such as the heat of its target, to find and destroy enemy aircraft.

Gyroscope A wheel with a spin axis that always tries to point in the same direction.

Hydraulic Using a control system that is operated by pumping liquid—such as oil or water—through pipes.

Instrument rating Qualification acquired after pilots train to fly using just their instruments so that they will be able to fly when visibility is poor or zero.

Launch vehicle Rocket or other form of launcher that propels a spacecraft into space.

Mass Amount of matter of which an object is made. On Earth, an object's weight is the same as its mass.

Meteorology Science that involves study of the atmosphere and weather.

Navigation Identifying and following a course.

Paradox Something that seems impossible or contradictory and goes against expectations.

Parafoil Also called wing parachute or ram air wing. High-performance parachute made of two layers of fabric joined by dividers that form a line of cells, which fill with air to make an airfoil.

Paratroops Military unit trained to parachute.

Pascal Unit of measurement of pressure that equals 1 newton per square meter.

Payload Load carried by a spacecraft or aircraft for delivery or for use on a mission.

Pitch Motion of an aircraft about its lateral axis that makes its nose tip up or down.

Propellant Something that propels (drives forward), such as the chemicals used to fuel a rocket engine.

Propulsion Act of propelling, or driving something forward.

Radar System that uses radio waves to detect and locate objects and movement.

Radiation Form of energy released in invisible waves and particles.

Reconnaissance Exploratory survey to gain information.

Relativity Relationship of one thing to another.

Rendezvous Meeting of two spacecraft in space, with or without docking.

Rocket booster Solid-fuel rocket that provides a launch vehicle with extra thrust.

Roll Motion of an aircraft about its longitudinal axis that makes it bank to one side or the other.

Rotor Turning part of a machine, such as the turning blades of a helicopter.

Rudder Movable part on the tail fin of an aircraft that is used to control yaw.

Solar cell Device that converts the energy from sunlight into electric energy.

Sonic boom Sound heard on the ground created by a shock wave from an aircraft flying at supersonic speeds.

Stabilator Also called all-moving tailplane or all-flying tailplane. Tilting tailplane that performs the combined jobs of the stabilizer and elevator.

Stage One of the launch sections of a rocket that is discarded after its fuel is used up.

Tail fin Vertical stabilizer placed at the rear of an airplane's fuselage, forming the upright part of the tail.

Tailplane Also called horizontal stabilizer. Horizontal part of an aircraft tail that sticks out on either side of the tail or rear fuselage to help keep the aircraft stable in the air.

Thrust Force that propels vehicles, such as spacecraft and aircraft, and is produced by propellers, jet engines, or rockets.

Transponder Radio or radar receiver that emits a signal when it receives a signal.

Trim Moving the center of pressure in an aircraft forward or backward until it lines up with the center of gravity, making the aircraft level.

Turbine engine Also called a gas turbine or just a turbine. Type of rotary engine in which whirling blades are rotated by pressure from gas or other fluid.

Turboprop Turbine engine that uses a jet of gas to spin a propeller.

Velocity Speed in a particular direction.

Waypoint Geographical location along a route, identified by longitude and latitude and used to define aviation routes.

Winglet Turned-up wing tip shaped to reduce drag.

VOLUME INDEX